Grapes

Grapes

Harry Baker and Ray Waite

MITCHELL BEAZLEY

THE ROYAL HORTICULTURAL SOCIETY

Grapes
Harry Baker and Ray Waite

First published in Great Britain in 2003 by Cassell Illustrated,
an imprint of Octopus Publishing Group Limited.
This edition published in 2008 in association with the Royal
Horticultural Society by Mitchell Beazley, an imprint of Octopus
Publishing Group Limited, 2–4 Heron Quays, London E14 4JP
An Hachette Livre UK Company
www.octopusbooks.co.uk

The publishers will be grateful for any information that will assist them
in keeping future editions up to date. Although all reasonable care has
been taken in the preparation of this book, neither the publishers nor
the author can accept any liability for any consequence arising from the
use thereof, or the information contained therein.

The authors have asserted their moral rights.

A CIP catalogue record for this book is available from the
British Library.

ISBN 978 1 84533 374 4

Commissioning Editor Camilla Stoddart
Designer Justin Hunt
Editor Robin Douglas-Withers
Illustrator Patrick Mulrey

Set in Bembo

Colour reproduction by Dot Gradations Ltd.
Printed and bound by Toppan in China

CONTENTS

6 INTRODUCTION

10 VINES UNDER GLASS

26 VINES IN POTS

34 THE GROUND VINERY

38 CULTIVARS FOR GROWING UNDER GLASS

44 VINES IN THE OPEN

64 CULTIVARS FOR GROWING OUTSIDE

72 PROPAGATION, PESTS AND DISEASES

86 FURTHER INFORMATION

90 INDEX

Previous page:
'Cascade', a
vigorous black
grape that ripens
in early October.

Introduction

The grape, *Vitis vinifera*, has been grown for many thousands of years and is generally thought to have originated in Asia Minor and the Caucasus. Wild forms in many colours still exist in Georgia, Armenia and Azerbaijan. About 5000BC it was introduced into Syria thence to Palestine and eventually into Egypt. Records from the ancient tombs mention the irrigated lands of the Nile yielded abundant wine and there are frequent references to the cultivation of the vine in Greek literature and in both the Old and New Testament.

With the conquest of the Middle East and Western Europe by the Romans, the vine spread further and eventually reached Britain, probably in the 1st century AD, though according to existing records of that period it was never grown by the Romans to any great extent. They considered our climate to be unsuitable and most of the wine consumed was imported. Nevertheless, since then, grapes for wine have been grown in Britain more or less continuously, though its popularity has risen and fallen according to the climate and the economics of the time. In the Domesday Book (1086AD) 38 vineyards were recorded and the Isle of Ely, under the jurisdiction of the Church, was noted for its good wine for hundreds of years. Again, in the Tudor period, for a while near sub-tropical weather was enjoyed and substantial vineyards were planted. Conversely there were periods of decline after the departure of the Romans, and again during the time of the Black Death. There was also a mini ice age during the 13th century.

'Seyval' is a good hybrid that is easily grown outside in Britain.

In modern times the grape out of doors is enjoying a revival. This started in the 1960s when pioneer viticulturalists, like the late Barrington Brock at Oxted, showed that grapes could be grown outside successfully, and that good wine could be made provided the right cultivars were chosen, good growing techniques adopted and the right site selected. Today there is a greater acreage outside than ever before and there is even a British Vineyard Association based at Saxmundham to look after the affairs of viticulturalists.

In contrast, the grape under glass has enjoyed more or less steady progress since its introduction. At first, in the Middle Ages, only royalty and the nobility could afford to grow them but as glass became cheaper and labour more plentiful, its culture increased, though by the 19th century it was still only within the province of the affluent. It was in the hands of Victorian and Edwardian gardeners that the culture of glasshouse grapes reached its zenith. The techniques and cultivars they developed still apply today. The difference now is that, thanks to relatively cheap modern structures, more efficient supply and control of heat, many more gardeners can enjoy growing grapes under glass very successfully. It is hoped that this book will help the reader to achieve the objective of growing grapes well, whether outside or under glass.

GLOSSARY

Cordon
A single main woody stem from which spurs and laterals arise. Alternatively called a rod.

Espalier
A vine trained with a vertical main stem carrying tiers of horizontal branches (rod) pruned on the rod and spur system.

Lateral
A young shoot arising directly from a spur or main stem.

Multiple Cordon
A vine consisting of more than one main stem.

Pinching back (or stopping)
To cut or nip out the growing tip of a shoot (lateral).

Rod
See cordon.

Rod and Spur Pruning
A system whereby the fruit-carrying branches arising from the main stem (rod) are kept short by winter and summer pruning.

Spur
A short branch system maintained by pruning carried out on the main stem (rod) of the vine.

Sub-Lateral
A shoot arising from the axil of a leaf on a lateral usually as a result of pinching back or stopping the lateral.

Tendrils
Twining stems which arise from the new growth enabling the vine to cling and climb.

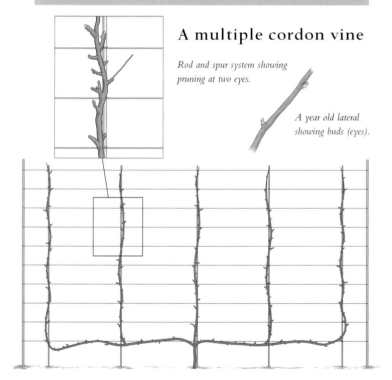

A multiple cordon vine

Rod and spur system showing pruning at two eyes.

A year old lateral showing buds (eyes).

VINES UNDER GLASS

During the Victorian heyday of the private estate and even to some extent into the early part of the 20th century, vines were grown in their own special greenhouses, called vineries. In some large privately owned gardens there were often separate vineries for early, mid-season and late fruiting cultivars. Invariably, it was more usual to divide the greenhouse into sections each with a different growing temperature.

After the Second World War, the days of the specialised vinery were over and relatively few people grew vines under glass. For some time now, however, interest has re-awakened and now increasing numbers of amateur gardeners are finding that they can grow a worthwhile crop of grapes in the roof space above other plants, even in a small unheated greenhouse especially in the warmer parts of the country. It is absolutely essential to ensure that the greenhouse, or for that matter the conservatory, has adequate ventilation both at the ridge and sides as air movement is absolutely vital for good grape culture. Growing vines in pots, although not a new system, can be even more convenient so that a small 2.4 x 1.8m (8 x 6ft) greenhouse can give good results.

In the British Isles, most dessert grapes need the protection of glass to ripen properly and achieve their full potential. Certain cultivars, notably the Muscats, need artificial heat in the spring, particularly at flowering to aid pollination, and also in the autumn, so that the grapes attain their best flavour. In view of the cost of a greenhouse, the cost of fuel and the amount of detailed work necessary to keep the vines under control, it is

'Mrs Pearson' sets easily and is a strong grower.

better to use the greenhouse for dessert grapes rather than wine grapes, which can be grown outside. The limiting factor in a small greenhouse is head room, but the current fashion is for a greater height to the eaves, so that problem may now be less significant.

The ideal structure for grapes is a lean-to greenhouse facing south; the roof rises across the whole width of the house affording as large an area as possible for training the vines, and in addition the rear wall collects a reserve of heat on a sunny day, which is released into the structure at night. However, vines can also be successful in free-standing span greenhouses.

Vines can be grown with or without heat. The latter provides a later crop, but limits the number of cultivars that can be successfully ripened. In a heated greenhouse the vine can be started into growth in mid-February, and ripe fruit expected in July and August, depending on the cultivar grown. In a cold greenhouse, growth will start naturally in March and April, with harvesting in late August and into September.

A vine grown without heat can be planted out of doors and the main stem (called the rod) trained through a hole made low down in the greenhouse wall, or it can be planted inside with the roots allowed to grow in the soil outside.

Vines grown in heated greenhouses are best planted in a specially constructed border. With roots inside, the grower has greater control and success or failure depends on his or her knowledge or judgement. The method is of particular advantage in the colder parts of the country where an early start is required or if there is any risk of the roots growing into areas that might become waterlogged in winter. In the past it was always said that for grapes started very early in the year the inside border ensured that the root action was on the move at the same time as the warmed top growth. However, this is a counsel of perfection and for most needs today outside planting will be satisfactory, making for easier initial preparation and less frequent attention to watering and feeding.

PREPARATION FOR PLANTING

Whether planting is to be done outside or inside the greenhouse, soil preparation means double digging, incorporating well-rotted manure or compost together with a dressing of general fertilizer such as Growmore. Bonfire ash can also be added if available. If there is any suspicion that the soil might become waterlogged, construction of a special bed will be necessary. Make the border 75–90cm (2½–3ft) deep with solidly constructed sides and base, for example brick retaining walls and a concrete floor. The floor should slope from back to front, and from one end to the other, and a land drainpipe 10cm (4in) in diameter be laid along the lower edge with an outlet at the lowest corner leading to a ditch, surface water drain, or a soak-away prepared by digging a large hole and filling it with stones or brick rubble. Spread similar drainage material to a depth of 15cm (6in) over the whole of the base before filling up the border space with soil. This soil should be capable of holding reserves of water and plant food, at the same time being open and porous. John Innes potting compost no.3 will provide a satisfactory growing medium; the recipe is 7 parts by volume of sterilized loam or fertile garden soil, 3 parts by volume of peat, 2 parts by volume of coarse sand, plus 300g (12 oz) of John Innes base fertilizer and 55g (2¼oz) of ground chalk to every 36 litres (1 bushel). If there is an objection to using peat, well-rotted farmyard manure or well-prepared mature oak and/or beech leaf mould will make adequate substitutes, but as these cannot be sterilized, some weed seed germination must be expected. If wood ash from a bonfire containing charcoal is available, this can also be included in the mixture.

The young plant will not require a very extensive root run in the first year, and the old practice of making up a narrow border for the first year and adding to the width annually has much to commend it, for it makes less initial work and reduces costs. This method is of particular importance for the larger traditional vinery. It ensures that a medium full of nutrients and in good physical condition is being added year by year during the early life of the vine. Concrete building blocks provide a convenient means of retaining the compost, but ordinary house bricks or planking will serve equally well.

The growing vine will need to be trained inside the greenhouse so preparation must include the fixing of wires for support. It is usual to run the wires along the length of the structure starting at about 90cm (3ft) off the ground, and continuing at about 22cm (9in) apart and 30–45cm (12–18in) from the glass.

In a small greenhouse it is only feasible to allow 30–35cm (12–14in) from the glass, but the young shoots tend to grow vertically and quite rapidly and will soon stub themselves against the glazing and become scorched or distorted if they do not have enough space. It is also essential to keep the air moving between the glass and foliage, so the greater the space given to the wires the better. In a larger greenhouse, allow about 45cm (18in) between the wires.

PLANTING

In the small greenhouse there will probably be room to plant only one vine. This is best done at the opposite end to the door so that the rod can be trained below the whole length of the ridge of the greenhouse. Alternatively, the vine can be planted to one side, at the end near a corner, and run along the side or be allowed to develop long enough so that it can ultimately be trained below the ridge. A main rod trained along the side can have vertical rods arising at intervals. In fact, vines are so accommodating that they can be planted in so many ways, provided that subsequent handling of growth and fruit is not made difficult. When planting outside it will be necessary to bring the young rod inside; this is best achieved in a 'glass to ground'

An established vine inside the greenhouse, with the main rod growing through a slot cut in a plywood board.

structure by replacing a bottom pane of glass with treated plywood or hardboard, or, for preference, a sheet of aluminium. A large slot will need to be cut at a convenient height and must be wide enough to allow for the rod to expand in the future.

However, traditionally vines were planted side by side and this method lends itself perfectly for the large greenhouse, especially of the lean-to design. Plant the rods at 1.2m (4ft) apart as the spread of lateral growth can be expected to reach 60cm (2ft) or more either side.

Young vines are usually sold in containers and therefore can be planted at any time of the year, but November and December, when the plants are dormant, are the best, as at that time the main stem can be cut back by two-thirds without fear of excessive sap bleeding. This severe pruning ensures vigorous growth in the summer and also stimulates the lower buds to form side shoots (laterals), so encouraging lower spur systems to be formed. If plants are not obtained until later in the winter, it is unwise to prune them, otherwise the rising sap will bleed for several days from the cut and considerably weaken the vine. Similarly, vines planted in full growth should not be pruned until after leaf fall.

A young vine viewed from outisde the greenhouse, growing through a slot in aluminium sheeting.

It is not always possibles to purchase strong vines (stem about 1cm (½in) in diameter) in which case it is better to pot on into a larger pot, reduce the main stem to one-third its original length, and allow the plant to grow for another season before planting in the following winter.

15

THE FIRST YEARS

In the first year after planting (and after the initial hard pruning) allow the main stem to grow unchecked so that it produces the maximum length of vigorous shoot. If the roots are growing well, the main shoot should reach 3m (10ft) or more and produce lateral shoots over much of its length. During the summer these shoots are stopped when they have made five or six leaves, and any sub-laterals (side growths formed on the laterals) produced are cut to one leaf. Both main stem and laterals are loosely tied into position on the wires as they grow, allowing the stems ample room to increase their girth.

At the end of the season and as soon as the leaves have fallen, the main stem is cut hard back again, removing two-thirds of the summer's growth. The lateral shoots are pruned to one plump bud.

In the following years the same technique is carried out until the main stem, after being pruned by two-thirds of its length, has reached the furthermost wires. Ultimately this form of training will result in what is known as the rod and spur system, a permanent framework consisting of a main stem with the laterals forming spurs at intervals of 22–30cm (9–12 in) along

The glasshouse vine, grown on the rod and spur system of pruning – before (left) and after (right) pruning immediately after the leaf fall.

it. It is from these spurs that the fruiting shoots arise each year. The main stem between the spurs is barren. This is the simplest form of training, but it is also possible to grow two or more rods on one plant or allow the rod to branch and extend until one vine fills an entire greenhouse, as does the famous vine at Hampton Court, London.

WINTER PRUNING OF ESTABLISHED VINES

Vines should be pruned immediately after leaf fall, which is usually in late November or December. At this time the greenhouse must be quite cold in order to keep the vines dormant. This is done by giving as much ventilation as possible after harvest so that the current year's growth is fully ripened. This hardening of the growth will, with the early winter pruning, avoid the real risk of a harmful level of bleeding when the sap rises in the spring. Bleeding can be prolonged and abundant, and although various 'cures' have been suggested, none are wholly satisfactory. If later pruning is unavoidable, the application of carpenter's knotting to the cuts immediately after pruning may help. Many different kinds of treatment have been recommended in the past but none are really effective once excessive bleeding commences. It is much better to avoid the trouble than to cure it as loss of sap might debilitate the vine.

With the established rod and spur system, winter pruning consists of cutting back all growth made in the past year (other than that required for extension or replacement) to within one or two buds of the older wood. One bud is enough if it is strong and plump, but often the basal bud is small and unlikely to give fruiting wood. Pruning to two buds will give two growths; should one become damaged or broken away, there is still one to be tied in to fill the space (see illustration, p.19).

Indoor vines can also be grown on a replacement pruning system. This requires more skill but will generally result in higher quality fruit. With this system the winter pruning is much the same as that for the Guyot method of growing vines in the open (see p.55). Two new main stems are allowed to grow each year, one of which is retrained for fruiting and the other cut back to two buds to produce the two rods for the following year. The rod that has fruited is cut out completely.

Winter Maintenance

The practice of scraping away the oldest bark from the main rod takes time and some growers regard it as unnecessary, but with the increasing reduction of available chemicals to amateurs and professionals alike, it is well worth finding the time to do the job. Special attention needs to be given to the flaky bark on the spur, with extra care being taken near the buds, which can be easily damaged. Much of the fibrous bark can be pulled or rubbed away, but when scraping with a knife the green tissue must not be exposed or cut; because of this it is wise to use an old blunt knife. By scraping away the old bark, over-wintering pests, such as greenhouse red spider mite and mealybug, are either removed or exposed to a pesticide, which can be sprayed on, but preferably brushed on ensuring that it is rubbed into crevices and angles of the bark.

When scraping the rods, the work is made much easier if they are untied and suspended from the wires. Remove the old ties, as these too can harbour pests. After scraping, the rods should be tied back loosely to the wires for the ease of subsequent work.

The next job is to renovate the inside bed. The remains of the old mulch are raked from the bed and the surface soil gently pricked over with a fork. This loose soil is raked or brushed off with a stiff broom or besom. The roots thus exposed are immediately top-dressed with a good loam-based compost such as John Innes no.3.

No more needs to be done for the next two or three months, until the vine is started into growth. As this time approaches, the soil in the bed is watered thoroughly; a second watering a week later may be needed if the root area has become very dry during the winter.

After watering, a mulch of well-rotted farmyard manure, spent mushroom compost or similar material should be applied; this will help to regulate the soil moisture during the growing season. This is also the time to extend the vine border in width as necessary.

For grapes planted outside the greenhouse, no special treatment is required other than an application of a good general fertilizer, supplemented with a dressing of sulphate of potash and liberal mulch of organic material as recommended for feeding outdoor grapes.

Removing a surplus shoot from a spur.

THE SPRING ROUTINE

If warmth is required for the needs of other plants, heat may be provided in February; the middle of the month is an ideal starting time for vines.

The rods that have been loosely tied to the wires for the winter operations are untied and lowered so that when supported about one-third of the way up by a long loop of stout garden twine, the end is bent down under its own weight, almost touching the ground. In this way, the rising sap is prevented from rushing to the top buds and stimulating them into growth at the expense of those lower down. When growth has started on all the spurs, the rods are tied into their summer positions on the wires.

Temperatures at this stage may rise rapidly in sunlight and as a gentle start is required, careful ventilation must be given at about 19°C (66°F). With artificial heat a night temperature of 4–7°C (39–45°F) should be the aim.

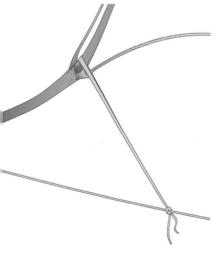

A lateral shoot being lowered with a running noose.

Keep the atmosphere moist by damping down the floor and walls on fine days. Until flowering begins, the rods can be sprayed overhead with clean water, but do this in the morning before the sun becomes too hot, for there can be a risk of scorching the young foliage by the action of the sun's rays through the water droplets.

If it is necessary to reduce the number of growths to give one shoot on a spur, the surplus ones should be removed as soon as possible. In practice, it is safer to pinch back the weaker unwanted shoots to two or three leaves, in case a potential fruiting shoot is accidentally broken when tying in (see illustration above).

As the shoots will grow vertically, they must be brought down to the wires; this can be rather tricky as they may easily snap at the junction with the older wood. Growths should be trained down gradually, doing this approximately every other day (see illustration p.19). Often, depending on the cultivar, it will be found that it is quite possible to readily tie the young growth straight away to the horizontal wire. It is unwise to tie in while the growth is very young and brittle, and better to wait until a little fibrous tissue has formed. The task should be done in the morning, for although later in the day the shoots can be limp and more easily managed, at night they become turgid again and if tied too tightly may break. Tendrils should also be removed before they are allowed to develop, not only to conserve the energy of the vine, but if the tendrils are allowed to remain, they can scramble into foliage and growth but, more importantly, become entangled in bunches of fruit. As the light intensity increases, apply light shading to the glass to prevent leaf scorch. This may need to be done as early as April with a small greenhouse. Shading will also help to reduce excessively high and too-rapid rises in temperatures and thus to some extent prevent conditions becoming too dry.

FLOWERING AND POLLINATION

By the time the shoots are 45–60cm (18–24in) long, usually at about five leaves, the flower trusses will be quite prominent and a start can be made on pinching out the shoot tips. Two leaves are allowed to form beyond the flower bunch before the growing tip is removed. If no bunch has been formed, the shoot is left to make six or eight leaves before pinching. All side shoots that form on the newly grown sub-laterals should be pinched

A stopped shoot showing the formation of sub-laterals.

at one leaf and any further shoots resulting from this pinching should again be pinched at one leaf. At flowering time, it is a good plan to dress the inside border with a generous layer of clean straw, which gives more water retention and helps to keep a drier atmosphere.

At this critical stage, a drier atmosphere is important. although paths can be damped down in mornings and afternoons when it is sunny and warm.

Although some cultivars set their fruit very readily, it is unwise to leave pollination to chance. With freely setting cultivars like 'Black Hamburgh', 'Foster's Seedling' and 'Alicante', all that is needed is to give a brisk shake to the rods. For shy setters like the Muscat types, the easiest effective method is to stroke the bunches gently with cupped hands in order to transfer the pollen from the stamens to the receptive stigmas. If pollen can be transferred from a different cultivar, so much the better. Always try to pollinate at midday. At this critical stage, keep the atmosphere on the dry side, and although warmth will help pollination, air must be kept circulating. Even on a cool day, a little ventilation will help. Do not spray overhead on such days.

THINNING THE CROP

If all the bunches are allowed to mature, the quality of the fruit will suffer; in addition, the vigour of the shoot growth is lost and fruiting may be upset for several years. The number of bunches left to fruit must therefore be regulated. As a general rule of thumb, a pound of fruit is retained for every foot run of rod (450g per 30cm). This lightening of the load should be done as soon as it is possible to assess which bunches show promise of being shapely, well set, of good size and in such a position that they will develop freely.

Next, the grapes must be thinned. It is not a difficult task, but it takes time. Special grape thinning scissors are obtainable but ordinary small scissors will do, provided they are sharp, particularly at the tip, and are not too cumbersome. A small stick about 15cm (6in) long, forked at one end, is helpful in steadying the bunch. Grapes should never be touched with the fingers, as this damages the bloom on the skin. This damage not

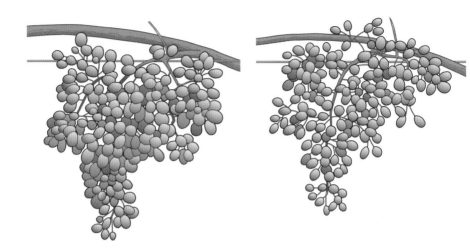

Above: bunches before (left) and after (right) first thinning
Below: bunches before (left) and after (right) second thinning

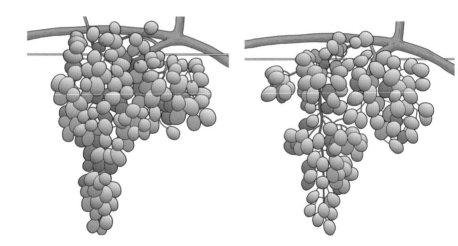

only disfigures the bunch, but removes the extra protection given by the water-repellent qualities of the bloom.

Occasionally it is necessary to shape the bunch before thinning, and this usually involves the removal of a straggling arm at the top of the bunch. Any large bunches, referred to as shoulders, are supported by looping them up to the wires with raffia. Excess grapes in the middle of the bunch are cut away first, then any undersized berries are removed as these are probably imperfectly set. Next, the outer berries are spaced to allow room for swelling. The shoulders should not be thinned too much as they should stand firm and square when the bunch is harvested.

Thinning can seldom be completed in one operation and it is usually necessary to remove a few more grapes three or four weeks after the first thinning. A final thinning may be required after the pips have formed when the grapes undergo a second period of swelling.

WATERING AND FEEDING

A vine in full leaf will make heavy demands on soil moisture in warm weather. The intervals at which water is required will vary according to the climate and soil conditions. If the vine's roots are confined inside the greenhouse, a generous application of water will probably be necessary every seven to ten days, and provided drainage is good, the risk of overwatering will be slight. It is better to water thoroughly so that moisture penetrates a couple of metres into the border soil. If the roots are outside, they may be supplied by natural rainfall for much of the year, but they should not be forgotten in prolonged dry spells.

Feeding will be required while growth is active. For good-quality fruit the main need is for a high-potash feed, and any of the tomato fertilizers available will be ideal. Water on a solution at two- to three-week intervals from about a month after starting into growth until the fruit begins to ripen.

Sometimes the vine is not vigorous – a condition brought about perhaps by overcropping, old age or poor soil conditions, in which case the application of a general feed, i.e. with equal quantities of nitrogen, phosphate and potassium, or one containing higher nitrogen will be beneficial, but these should only be used from April to May.

RIPENING

Throughout the summer and autumn a healthy vine will grow vigorously. Any new shoots not needed to fill gaps in the foliage canopy or to add to the length of the rod for the following year are cut back after one new leaf has formed; this must be done promptly or the growths will rapidly get out of control. Remove tendrils immediately as they form.

When the grapes start to change colour, there is a greater risk of splitting. This stage is easily recognized in black or tawny grapes, but it is not so apparent with green ones, although a slight change of colour towards white or amber is noticeable.

A damp atmosphere encourages splitting. Damping down of the path and other surfaces should be done by the middle of the day so that any excess moisture has had a chance to dissipate before night. If there are other plants in the house, they should also be watered well before evening. Leaving the lee side top ventilator open by about 3cm (1¼in) will help to dry the atmosphere and will also control, to some extent, any early build up of heat in the morning. The air temperature should never be allowed to rise rapidly, because the temperature of the grapes rises slowly, and moisture from the air will condense on the fruits, which may then split. If the soil is allowed to get very dry, the fruit skins will harden and when water is finally given, they are likely to split. A thin mulch of straw applied over the soil when the fruit has formed will help to keep the soil uniformly moist and the air in the greenhouse on the dry side. From the early stages of ripening, look over the bunches two or three times a week for diseased grapes, removing those with a split skin; continue this until the end of the season. Cut out the bad grapes using scissors and the forked stick, working with care so as to keep the bloom intact.

Grapes are not wholly ripe when fully coloured and require 'finishing'. This means allowing them to hang for a period on the vine so that sugars are formed. The finishing period varies with the cultivar and time of year. 'Black Hamburgh' and 'Foster's Seedling', ripening in the summer, need only two or three weeks to mature fully; 'Muscat of Alexandria' and 'Mrs Pince', which ripen later, need to hang for eight to ten weeks before they are at their best. When harvesting the grapes, handling will be made easy if the branch carrying the fruit is cut with a 'handle', i.e. leaving about 5cm (2in) on either side of its joint with the main stem. This handle allows the bunch to be inspected, carried, mounted on a showboard or placed on a dish without disfiguring the fruit.

A cut bunch of 'Muscat of Alexandria' attached to a display board for exhihibiton.

Ripe grapes may be kept in good condition on the vine for several weeks by using artificial heat and ventilation to maintain a cool dry atmosphere (about 7°C/45°F).

If a cool, even temperature can be provided in a room, it is worth cutting the late grapes with a stem in December. The stem can be placed in a jar of water and in this way fruit will keep for a considerable time. 'Lady Downe's Seedling' is particularly good for this purpose.

VINES IN POTS

The cultivation of vines in pots has several advantages. The containers can be readily moved within the greenhouse and moved outside after fruiting. Such outdoor treatment will ripen the shoots and give the cold environment necessary for satisfactory dormancy. Clay pots will need to be plunged to reduce the chance of frost damage to the containers; plunging also gives stability during windy weather whatever the receptacle.

Much of the work described for vines under glass applies equally to the management of pot grown grapes except that they are invariably trained as standards, with a long stem producing a 'head' of shoots (see illustration overleaf). The method deserves to be more widely used because worthwhile crops can be obtained, especially in the smaller greenhouse. Where space is limited, more than one cultivar can be grown and, with care, the more usual crops can be cultivated at the same time. For example, the vines can be taken out of the greenhouse in the autumn, leaving room to house chrysanthemums. Heating will give a long season of growth and therefore earlier ripening, but an unheated greenhouse can also give good results.

Early fruiting cultivars will be found to be the most satisfactory with 'Black Hamburgh' still possibly the most suitable; but 'Buckland Sweetwater', 'Foster's Seedling' and the Frontignan types all give good results.

Well-established rods should be obtained and are best potted just as the roots become active in the spring. Weak rods will need to be grown on to build up the main stem and it will be

'Black Hamburgh' has a good flavour and does well in containers.

26

New shoots arising from a pruned main stem.

necessary to cut the rod back to one bud to induce more vigorous growth during summer. Any pruning should be done during November and December. Rods that have a diameter of 1cm (½in) or more can be pruned to a bud (eye) at 1–4m (3–5ft) above the pot. The length of the stem chosen will depend on the height of the greenhouse.

Vines may need to be potted into a larger size pot as the root system grows, and this is best done as the roots start into active growth in the spring. A vine in a 13cm (5in) pot should be potted on into an 18cm (7in) pot, but plants with larger rootballs can be moved into a 22cm (9in) pot. A loam-based compost, e.g. John Innes no. 3, is best, because it will ensure that the plant will not become top heavy.

Once started into growth, the shoots soon lengthen and the top four or five shoots should be left to develop a 'head', being pinched when they have made five or six leaves. Any growths on the stem should be removed completely. Flower clusters are unlikely to appear but if they do, remove them, as they should not be allowed to develop at this early stage.

For the second year the ripened shoots should be pruned in the dormant season to one bud. This is

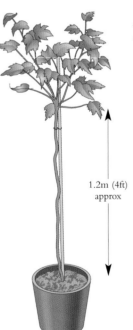

A pot-grown vine trained as a standard.

1.2m (4ft) approx

the start of building up the spur systems. These will form the permanent head of a pot-grown vine. As active growth begins in the spring, scrape off the existing top layer of soil then add a top dressing with John Innes potting compost no. 3 to the top of the pot, being careful to leave enough depth for thorough watering. To encourage strong growth and to ensure that it is well ripened, stand the vine out of doors from midsummer onwards.

Branches are secured to the central cane with raffia or garden twine.

During the third year, prune each shoot hard to one good bud in early winter. A move into a larger container will almost certainly be required, but this is likely to be the last move as a vine can remain in the same pot for many years with top dressing and feeding.

The vine can produce its first fruit in the third year. Pinch the growths at two leaves beyond the bunch, and any laterals and sub-laterals at one leaf.

The growths will need support and this can be done by tying them to a central cane in a maypole fashion or to a pre-formed 'umbrella' of wires or twine, such as can be used for training standard roses. A general liquid fertilizer is best applied every seven days up to fruit set, after which a high potash feed (e.g. a tomato fertilizer) should be used.

In warm situations, it may be possible to finish ripening the fruit outside, but it will need to be protected from bird and wasp damage, by covering with netting, or enclosing the bunches in nylon stockings (see p.74–79).

Although a vine grown as a standard is a convenient method of training, a simple rod and spur system can also be adopted. In this case the lower shoots may be allowed to develop and should then be pruned back in subsequent years, as described for permanently planted vines (see p.17).

Several novel methods of growing grapes in relatively small pots can be adopted and adapted using techniques perfected by Victorian and Edwardian gardeners, which are well within the scope of the keen amateur. By growing vines in small pots, bunches of grapes can be brought into the house and displayed still attached to the plant, providing fresh fruit and an interesting talking point.

There are two methods of training that can be used. For the best results, it is necessary to use growth from a reasonably well-established vine, when shoots produced from one of the lower spurs should be allowed to grow as long as possible – which can be as much as two metres or more. It is essential that the one-year-old rods are ripened and that only the section with thin brown bark is utilized.

The first method produces several bunches on a fairly short stem. In spring, the young rod is shortened for ease of manipulation; the tip is inserted from below through the drainage hole in a 20cm (8in) or 23cm (9in) clay pot, allowing three or four mature eyes (buds) above the rim. A loam-based potting compost is then firmed round the rod and the pot filled to 2.5cm (1in) below the rim. At this stage it should be possible to insert an upright stout bamboo cane, which is shortened to approximately 15cm (6in), above the tip of the inserted rod. Once the parent vine starts growing, the compost needs to be

Below left: Rod inserted through a hole at the base of a 20cm (8in) clay pot.

Below: The same pot having been placed in position.

Above: The pot ready for severing. Note the ripe fruit of 'Black Hamburgh'.

Above right: The free-standing pot, severed from the rod

kept moist and the rod sprayed with clear water as the buds also break into leaf. It will not be too long before a root system is formed, so within two months a weekly high potash liquid feed can be safely applied. Provided the buds have the propensity for flowering, pollination and thinning the fruit takes place as described earlier, in the normal way. When the grapes are ripe the rod is separated from the spur with a clean cut that is level with the base of the pot.

As an alternative, a similar young rod can be laid horizontally so that buds are positioned over the centre of a 15cm (6in) clay or plastic flower pot that has already been filled with potting compost. It is important that the compost forms a slight dome in the centre as the bud must come into close contact with the surface to facilitate root production. For ease of handling, the pots are best positioned on a length of slatted staging and the rod tied securely down to further ensure adequate contact with

Left: The rod layered on pots. Note that the pots have been well-filled.

the compost. As roots form, the bud makes its usual stem growth, which invariably produces a single bunch of grapes. Again, the compost must be kept moist, especially as the pots, out of necessity, have been over-filled. Feeding as prescribed previously must also continue. Once the fruit is ripe, individual pots can be cut away and brought into the home.

Right: Severing the individual pots.

Below: A layered rod taking root.

THE GROUND VINERY

The ground vinery is an excellent way of growing a cold glasshouse cultivar in a situation where a full-sized glasshouse is neither wanted nor practical. It is relatively inexpensive to construct, hence its Victorian name 'The Curate's Vinery'.

The vinery should be sited in full sun and ideally orientated east-west, though this is not critical, provided that the site is sunny and sheltered. The structure is glazed either with glass or rigid, clear or translucent plastic that does not have a UV inhibitor. Glass provides more heat and is, therefore, preferable, especially in the north. The one illustrated, at the RHS Garden, Wisley, in Surrey, is glazed with plastic. One side at least should be hinged (the south side if placed east-west) to open fully for ventilation and cultural operations.

The vinery shown measures 2.25m (7ft 6in) long, 90cm (3ft) wide, and 50cm (20in) high at the centre. It should be noted that the length depends upon the glazing used. Horticultural glass usually measures 90 x 60cm (3 x 2ft) and if this is used, the length would be 210cm (6ft 6in) with 5 x 2.5cm (2 x 1in) wooden framing. A better length for a heavier yield is 2.25m (7ft 6in) using 105 x 60cm (3½ x 2ft) glazing. As glass is heavier than plastic, two lights per side are necessary (see diagram). The ends are boarded and it is recommended that a 10cm (4in) gap should be left at the apex of each end board for top ventilation.

The floor of the vinery should be solid. The one shown stands upon two rows of 45cm (18in) square paving slabs, and they are painted black to improve heat absorption.

The Ground Vinery at the RHS Garden, Wisley, in autumn, with one side open.

Slates, bricks or stone are alternative materials. The vinery is raised above the floor on a plinth of bricks, leaving half a brick space between each brick for basal ventilation. Finally, the structure should be securely anchored to ensure it does not move in strong winds.

The ground vinery at RHS Wisley. Note the vine, planted outside, is trained through an aperture in the end board.

A ground vinery glazed with glass. Note the 10cm (4in) gap between the ridge and end board for ventilation.

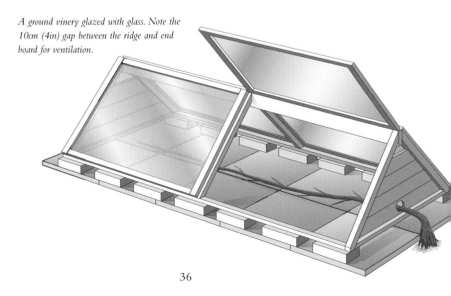

Cultivation: the vine is grown as a single cordon trained horizontally along a wire. The wire is fixed centrally to the end boards at 15cm (6in) above the ground (see diagram).

Planting: Prepare the planting site well, in the same way as described for a vine in the open (see p.48). The soil should be friable, weed-free and well drained. Plant the vine outside, about 15cm (6in) away from the centre of one end board. Lead it under the board or through an aperture and tie the plant to the wire.

Pruning: The vine is trained and pruned on the rod and spur system in the same way as a grape in a glasshouse (see p.16). The fruit carrying laterals, 25–30cm (10–12in) apart, are taken at right angles to the rod and the bunches lie upon the ground, hence the need for a clean, solid floor. The floor provides extra warmth so that the grapes ripen well and early. Light shading on the south side will be necessary from late spring onwards, as well as extra ventilation by raising the lights on very hot days.

Recommended cultivars: 'Black Hamburgh', 'Buckland Sweetwater', 'Foster's Seedling' (see pp.38–43 for details).

Cultivars for growing under glass (including pots)

Greenhouse grapes can be grouped as follows:

SWEETWATER (S)

These are early grapes, quick to mature and ripen. They are ideally suited for an unheated greenhouse and are very sweet and juicy but without the muscat flavour. They have very thin skins when at their best but do not hang on the vine in good condition for long.

MUSCATS (M)

The finest grapes for flavour, but they need heat. They are the second group to mature and they hang well in good condition if a little warmth is provided. The fruit is firm and luscious and if allowed to hang until the skin starts to wrinkle, the berries often take on the flavour of raisins. They are unfortunately difficult to get to set well unless they have warmth and some hand pollination.

VINOUS (V)

These are the really late grapes and in the past were grown to provide fruit well into the new year. They usually grow strongly and crop well but need to hang for a long time in a warm greenhouse to develop their best flavour. They are not a worthwhile proposition unless heat can be provided for them in the early winter months.

'Lady Hutt' is a dessert grape that makes a very large bunch.

The figures given are a guide to a reasonable number of bunches for a healthy vine, per 3.6m (12ft) of rod. A medium-sized bunch weighs about 450g (1lb)

Alicante (V)

A black (mid- to late season) grape. Vigorous grower. Fruit sets freely and needs early and severe thinning. It is somewhat inclined to form an unshapely bunch but when well thinned it can produce a handsome and imposing bunch for show with its large, almost round, jet black berries covered in a heavy blue/grey bloom. 9 bunches.

'Alicante' is a good exhibition cultivar.

Black Hamburgh (S)

This has the best flavour of this group and is justifiably the grape best known and most widely grown under glass in this country. Excellent for pots. Of good constitution, it sets fruit freely and can be ripened well in an unheated greenhouse. If allowed to hang for too long after ripening, the skin becomes very thin and easily broken and the fruit deteriorates. 12 bunches.

Buckland Sweetwater (S)

A round white grape carried in short, broad bunches. It sets freely and ripens early, particularly as a pot-grown vine. It has a pleasant flavour when in good condition but the skin thickens and the fruit deteriorates if it is allowed to hang too long on the vine. It is not a vigorous cultivar and needs to be fed well for good results. 12 bunches.

Canon Hall Muscat (M)

An almost round white grape that becomes pale amber on ripening. Of excellent flavour when well grown, but it is difficult unless heat is provided at flowering time and for ripening the fruit. It is possible to buy it grafted onto 'Black Hamburgh', which ensures stronger growth and a more manageable plant. Berries very large. 10 bunches.

Foster's Seedling (S)

The best of the white sweetwater grapes. Useful in pots. It sets freely, the bunches are of medium size and shapely, the grapes oval, juicy and of a pleasant flavour, ripening early. They should be eaten fairly soon after ripening or they will lose flavour. It has a good constitution and will grow well in an unheated greenhouse. 12 bunches.

Frontignan (M)

Both the black and the white 'Frontignan' grapes are early maturing with small grapes of good quality. Well suited to growing in pots under glass. 14 bunches.

Golden Chasselas (S)

A small, round white grape giving many small bunches of sweet fruits, ripening early. Good for pots. 20 bunches.

Golden Queen (S)

A very good-looking grape of strong constitution and ease of pollination. The bunches are of medium size and even triangular shape. The flavour is reasonable with fruit ripening mid-season. 12 bunches.

Gros Colmar (V)

A strong-growing, round black grape with very large fruits and handsome bunches. The skin is thick and the flavour poor until the fruit has had a long period of ripening in a warm greenhouse through the early winter. 10 bunches.

Lady Downe's Seedling (V)

One of the better-flavoured vinous grapes with round black berries carried in long tapering bunches. It is of good constitution but requires warmth at flowering time to ensure a good set of berries, and in the early winter to ripen its fruit. 10 bunches.

Lady Hutt (S)

A round white grape ripening several weeks later than others in that group. It is a robust grower, sets freely and produces large bunches of thin-skinned juicy berries. 5 bunches.

Madresfield Court (M)

An early ripening, oval black grape of high quality. Its berries are large and covered with dense blue/grey bloom, the flesh firm but juicy and of good flavour, but the skin is tough. The berries are liable to split on ripening and for this reason some growers used to leave top shoots to run on without stopping at ripening to reduce sap pressure. 10 bunches.

Mrs Pearson (M)

A round white grape that ripens late. It grows strongly, sets freely, bears fruit of excellent flavour but with thick skins and will hang in good condition after ripening. It does, however, need warmth throughout the ripening and holding period. 10 bunches. (See p.4).

Mrs Pince (M)

A strong-growing, oval black grape that produces late grapes of excellent quality, but it requires warmth and hand pollination at flowering to obtain a good set of fruits. It also requires warmth throughout the early winter when it is ripening. 10 bunches.

'Muscat of Alexandria' is the best flavoured of all greenhouse grapes, and can produce long, tapering bunches.

Muscat of Alexandria (M)

When well grown, there is no better-flavoured grape. It has a good constitution but needs warmth and hand pollination if it is to set a good crop of oval white grapes. Warmth is also needed to ripen the fruit, although if heat is used in the greenhouse in spring to start the vine into growth early the grapes should ripen in the September sunshine. 10 bunches.

Muscat Champion (M)

A round, large red grape that does not keep in good quality long after ripening, but otherwise first class.

Muscat Hamburgh (M)

An oval black grape of excellent quality that ripens before most of the other muscats. It is sometimes recommended for growing in an unheated house, but it is difficult to grow well. It needs warmth and hand pollination at flowering time (cross pollination, using pollen from a different cultivar is helpful). Although it is fairly vigorous in growth, it has a strong tendency to shanking (see p. 84). 9 bunches.

New York Muscat (M)

A good-flavoured cultivar that seems to do well in the north of the British Isles provided protection can be given. 15 bunches.

Red Flame Seedless (S)

An attractive grape that is often seen in supermarkets and fruiterers when imported in season. It ripens early, taking on a rosy red colour, and has a reasonable flavour. 12 bunches.

Thompson Seedless (S)

As this grape tends to be fairly late ripening it may be necessary to give it some heat, especially in colder areas. The berries are sweet and juicy. 12 bunches.

VINES IN THE OPEN

The grape, *Vitis vinifera*, is thought to have originated in Asia Minor and the Caucasus region. It is a sun-loving deciduous plant, which does best in temperate regions where frost-free springs and warm, dry summers are experienced, with winters cold enough to induce dormancy, but not so severe as to damage or kill the plant.

The United Kingdom does not often enjoy such weather, but it is possible to grow the grape outside with reasonable success, at least in the southern half and the Midlands, given the right conditions and provided the correct cultivars are planted. In general, outdoor grapes can be grown south of a line from The Wash in the east to south Wales in the west. They will not do well in areas of high summer rainfall, strong winds or where the climate is cool throughout the growing season. In the extreme west, therefore, and the north, it is best to grow the vine on a warm wall or solid fence, or to provide some kind of protection. Even in the south, the use of glass or plastic to protect against frost, and as a means of increasing the temperature, is very beneficial. In the West Country, for example, the growing of wine grapes in 'walk in' polythene tunnels has proved extremely worthwhile.

In essence, the United Kingdom is a marginal area for viticulture, and while there will be good years for the winemaker, there will be others when there are hardly any grapes at all or the wine at best might be regarded as 'vin ordinaire'. Remember that it takes two good consecutive summers to produce a full crop. The first is needed to ripen the

'Müller Thurgau' is a classic wine grape that gives an excellent wine.

44

wood and promote fruit bud development, and the second to obtain adequate fruit set and subsequently well-ripened, fully developed bunches.

GROWING ON WALLS, WOODEN FENCES, PERGOLAS AND ARCHES

The warmer and sunnier the position, the better the quality of the fruit in terms of flavour and sugar content and undoubtedly one of the best ways of growing grapes outside in this country is against a stone or brick wall with a southerly or westerly aspect. They can also be grown against a wooden fence. The extra warmth and shelter the structure provides makes it possible to grow grapes in cooler parts, where they could not otherwise be contemplated, and in the south, certain outdoor cultivars that need more heat can also be grown. However, it must be noted that soil at the base of a wall can become very dry, due to the absorbent nature of the brick or stone and because the wall can cast a rain shadow over the ground in its vicinity. This means that care must be taken in soil preparation (see p.48) and after planting to ensure that the vine is never stressed through lack of water. In addition, a wall-trained vine is more at risk from mildew and red spider mite owing to the hot, dry conditions.

Training methods

The vine is a flexible climbing plant and there are many ways in which it can be grown – for example, as a cordon in single or multiple form, as an espalier, as a combination of both, or even as a fan. Whichever method is used, the main point to bear in mind in training the plant is that enough space must be left between one framework branch (rod) and the next for the young laterals carrying fruit in the summer. Vertical rods should be spaced 1–1.2m (3½–4ft) and horizontal ones 45cm (18in) apart. Basically, they are pruned on the rod and spur system in the same way as the grape grown under glass (see p.10), except that the work carried out in the growing season comes later. A low wooden fence on a boundary, for instance, can be covered with grapes grown on the rod and spur system or the vines can be grown on the Guyot method (see p.55). A pergola or arch is best covered by growing the grape as a single or multiple cordon.

Top: A vine in winter trained as a 5-rod multiple cordon.

Right: The same multiple cordon in the spring.

Support system

A well-trained grape will require a supporting system of horizontal wires starting 45cm (18in) from the ground and spaced three brick courses apart, about 25cm (10in). These are secured to the wall with 10cm (4in) lead vine eyes driven into the vertical joints between the bricks every 90cm (36in) or so. Use gauge 16 (1.6mm) galvanised or strong plastic-covered wire. A wooden fence can be wired in a similar way. A pergola will need a system of horizontal wires at the top, 25–30cm (10–12in) apart.

Preparation and planting

Plant in the dormant season, ideally in November while the soil is still warm, or in March when the worst of the cold weather is over. A container-grown vine can also be planted in the growing season.

Prepare the ground thoroughly. As well as being dry, the soil at the base of a wall is often poor, containing builders' rubble, for example. If necessary, replace the soil with a good-quality medium loam over an area approximately 90 x 45 x 45cm (36 x 18 x 18in) deep. To improve soil moisture retentiveness, incorporate thoroughly a bulky organic material, such as well-rotted manure or compost, about two bucketfuls should suffice. Also fork in 110g (4oz) of a balanced fertilizer, such as Growmore, and 225g (8oz) of bonemeal. The same preparation applies equally well to a vine against a fence or pergola. Gardeners who prefer purely organic fertilizers can use seaweed meal in place of Growmore.

Plant the vine to a 1.8m (6ft) cane, 15cm (6in) away from the structure and to the same depth as it was in the nursery. Spread the roots outwards away from the wall and firm the soil during the filling-in process. A container-grown plant in leaf should be watered thoroughly before it is taken out of its pot. Do not disturb the rootball except to gently tease out the perimeter roots. When planting is complete, mulch the vine generously with bark, compost or well rotted manure.

Training and pruning

If a single cordon (rod) is all that is required, the strongest shoot on the young plant should be trained up the cane throughout the summer and any other shoots kept short by pinching them back to five leaves as and when necessary. Side shoots within 30cm (12in) of the ground should be removed altogether in late summer or in early winter.

To create a multiple cordon, the first task is to grow two strong vertical shoots emanating from the main stem at about 30–38cm (12–15in) from the ground. These are laid horizontally and tied to the lowest wire at the end of the summer and from them in future years will arise the rods of the multiple cordon (see photograph p.47). It may be possible to establish the two shoots in the first season if the plant is strong. If weak, it is best to grow it as a single cordon in the first summer and then cut it down to 30cm (12in) in November, ensuring that there remain two or three good buds at or near this point. In the second summer train the two rods vertically, keeping all other shoots pinched back to two or three leaves and then tie the rods horizontally at leaf fall. If a double, rather than a multiple cordon is all that is wanted, keep the rods trained vertically but space them 1–1.2m (3½–4ft) apart. Thereafter, treat them as described below.

The grape cultivar 'Madeleine Angevine' at harvest time.

Subsequent pruning and training

Throughout the growing season train and tie in the rod(s) as and when necessary to fill in the available wall space. In November, after leaf fall, any immature and unripened growth at the end of the leaders of each rod should be removed by cutting back to a bud on mature, nut brown wood. This is repeated each autumn until the required length of rod has been reached, side shoots (previous summer's growth) are cut back to one good bud near the base. This is done to form spurs along the length of the rod(s).

Pruning the established vine

Pruning, deshooting, pinching and tying in must be done regularly. The method is the same as for vines under glass (see p.19), except that the summer's operations are timed later.

(a) Summer

The first task is to thin out the young laterals growing from the spurs, but not until the flowers are produced. The shoots are then thinned to one per spur or one about every 25cm (10in). Retain those that are carrying the strongest flowers. The unwanted laterals are cut back to one leaf, rather than being removed completely, to reduce the risk of blind spurs. The retained laterals are stopped by pinching out the growing point at two leaves beyond the flower truss. Any sub-laterals subsequently produced are stopped at one leaf. Any barren laterals are thinned in the same way and stopped at six leaves, and their sub-laterals to one leaf.

(b) Winter

As soon as the leaves have dropped and before the end of December, the spurs are pruned by cutting back the previous summer's laterals to one or two buds ensuring the cut is made to a plump bud. Such a bud is more likely to produce a shoot with flowers in the next summer. Thin the spurs if necessary to 22–25cm (9–10in) apart.

The opportunity should also be taken to fill in empty wall space with extension growth or suitably placed laterals. An old rod carrying too many dead spurs can be replaced by leaving in a strong replacement. Tie in new growth where necessary and check any old ties to ensure that there is no constriction taking place.

Spur pruning a cordon in early winter.

Remove any thin, unripe extension growth by cutting back to strong, nut brown wood. The wirework should also be examined, renewed and extended if necessary.

Watering

A vine growing against a fence or wall outside needs very little watering once it is well established, except in periods of drought. While it is young, however, watering is advisable in the summer in dry periods to promote strong growth, which will be better able to withstand the ensuing winter frosts. A vine growing against a house wall where the soil is usually sheltered by the eaves of the house can become very dry at the roots. Grapes grown in hot, dry conditions are more susceptible to mildew. It is essential to keep the soil reasonably moist throughout the summer by irrigation, when necessary on such sites. This involves applying about an inch of water (17 litres over 0.8m²/ 4½ gallons over 1 sq yard) every seven to ten days, in hot dry weather. A mulch of well-rotted stable manure or compost applied in late February will help to conserve soil moisture.

A row of cordon vines after pruning in winter.

Feeding

In late February rake away the old mulch and apply Growmore fertilizer (7%N: 7%P$_2$O$_5$: 7%K$_2$O) or its equivalent at 66g/m^2 (2oz/sq yd) plus sulphate of potash 15g/m^2 (at ½oz/sq yd) over the rooting area and then apply a new mulch of well-rotted stable manure or compost to a depth of about 5–7cm (2–3in). Organic gardeners can use seaweed meal instead of the Growmore and sulphate of potash.

If magnesium deficiency symptoms appear, apply magnesium sulphate on a regular basis (see p.83). With dessert grapes give the vine a liquid feed high in potassium, about 2 litres (½ gallon) per plant every ten days throughout the growing season, but stop once the berries start to ripen.

Cropping and thinning: bunches and berries

The number of bunches to allow depends upon the age and condition of the vine. As a guide, a well-grown, mature vine should be able to carry one bunch every 25–30cm (10–12in) along the length of the rods. Certain cultivars that produce very small bunches with tightly clustered berries, 'Brant' and 'Cascade' for example, can be allowed more, at every 20cm (8in) or so.

Over-cropping must be avoided. A young vine in particular will frequently remain barren for a number of years after it has been allowed to carry too many bunches. Only two or three bunches should be left on a three-year-old vine, perhaps four or five in the following year, and so on, provided growth is sufficiently vigorous to sustain the crop. Overcropping and the removal of too many leaves in the summer will also depress colour, sugar content and flavour.

The thinning of the berries within the bunch (see p.22) is not necessary for wine, but it could be done to advantage for dessert grapes, provided they are cultivars that can produce large-sized berries, for example 'Siegerrebe', 'New York Muscat', 'Précoce de Malingre' and the Chasselas clones. Do not thin 'Brant' and similar cultivars as the berries are naturally small.

Growing grapes in a vineyard

Choosing the site

With vines grown in the open, the choice of site is most important. It must be sheltered and yet in full sun and at an elevation of not more than 90m (300ft). The ideal site is a south or southwesterly facing slope but level ground, or even a site sloping slightly towards the east or north is satisfactory. Avoid planting in a frost pocket as frost in the late spring can damage the blossom and young growth, possibly wiping out the potential crop for that summer. In situations where there is no alternative, the gardener should be prepared to protect the vines in some way, for example by covering them with hessian when spring frosts are expected.

Soils, preparation and planting

Vines are tolerant of a wide range of soils provided they are at least 30cm (12in) deep and well drained. Light to medium loams are ideal but very fertile land might cause problems with excessive vigour, and extremely shallow soils over chalk might cause lime-induced chlorosis and poor growth in very dry seasons. Very acid soils require to be limed to raise the pH to between 6.5 and 7.0.

Grapes root deeply and will not tolerate badly drained land nor a hard, impenetrable pan beneath the surface. Heavy clay, where there are drainage difficulties, should be avoided, but if this is not possible then some kind of drainage system must be installed as the first stage of soil preparation. The type depends upon the scale of planting. In a small area, where only one or two vines are planted, a simple soak-away 60 x 75cm (24 x 32in) made of broken bricks, clinker and rubble would suffice, but for larger areas, say for one or more rows of vines, a length of tile or plastic drains or even a herringbone system of drains may be necessary.

The planting site should be prepared well beforehand by double digging to break up any hard layers. Ensure the area is free of weeds, and add lime if necessary. Too rich a soil is undesirable and no bulky organic manure is required unless the ground is poor. Even then only a light dressing of well-rotted manure or compost should be incorporated. In the final preparations, and just before planting, rake in a balanced fertilizer, such as Growmore at the rate of 100g/m² (3½oz/sq yd), or the organic equivalent.

Plant dormant vines either in the autumn or the spring and not in the depth of winter to avoid any risk of damage by cold. In any event, weak vines with stems of less than pencil thickness should be over-wintered in a cold frame or glasshouse and planted out when the danger of frost is over. Most plants supplied by a nursery will be on their own roots and are perfectly acceptable, although a number of plants are being imported from the continent that are grafted on to rootstocks and these are usually available in the spring.

Grafted vines are resistant to the vine louse, *Phylloxera vastatrix*, although fortunately it is very rarely a problem in this country. Some rootstocks are more suited to certain soil types than others (see table).

Rootstock suitability for soil types

Shallow poor, stony dry	Deep fertile without chalk	Deep fertile with chalk	Heavy clay with chalk	Heavy clay	Chalk
5BB 125AA 3309	5C SO4 125AA	5C SO4 5BB 125AA	5C	5BB SO4	5C

Firm planting is essential and the roots of young vines (often pot grown) must be carefully spread out to encourage quick establishment. Subsequently, mulch them with a little rotted manure or compost so that the lower buds are covered for the winter. With grafted plants, the graft union should always be covered in winter. In the spring, the mulch should be pulled away from the stem. The graft union, indicated by a slight bulge, must be about 2.5cm (1in) above the soil to prevent scion rooting; if this occurs the desired effect of the rootstock on the vine will be lost and the plant will no longer be resistant to *Phylloxera*.

Training methods

There are many methods by which a vine can be trained in the open. It could, for example, be grown in a sunny corner as a vertical cordon, trained up a stout cane, or as a line of vertical cordons supported by canes and wires, or even as a clean-stemmed standard like a weeping rose (see photo, p.73). Cordons are particularly suitable where a relatively large number of cultivars, such as a collection, are to be grown. The cordons should be spaced 1–1.2m (3–4ft) apart. It could also be grown in a container (see p. 26–29). The grape in all of these examples would be spur pruned in the winter and the laterals pinched back to three or four leaves beyond the bunch in the summer to keep the plant in shape.

THE GUYOT SYSTEM

For vineyards large and small, however, the most widely used method of training is the Guyot system in single or double form with the vines planted in rows. This is a replacement system of pruning, whereby the wood that has fruited in the previous summer is cut out each winter and new wood tied in to fruit in the next summer. The Single Guyot has one fruiting arm whereas the more popular Double Guyot has two. The latter is preferred by most growers as it takes fewer plants for the same amount of space and production is about the same. There is a case for the Single Guyot where there is no room for two arms, where the soil is very poor or where a large number of cultivars are wanted relative to the land available. In either form, the arms are trained fairly close to the ground to take full advantage of the reflected and radiated warmth from the soil. Preferably the rows should run north to south to reduce mutual shading.

Support

Vines grown in rows in the open require a stout fence of posts and wires. For the Guyot system, the posts should be spaced 4–4.5m (13–15ft) apart depending upon the spacing of the plants within the row.

The end posts and the intermediates should be 2m (6ft) long, by 10cm (4in) diameter, driven 45cm (18in) into the ground.

The wood must be preserved against rot or resistant to it, preferably impregnated under pressure with a preservative or dipped. The fence should be strutted at each end. Galvanized fencing wire is used for the wirework. The two lowest wires are single and the three sets of upper wires are double. Secure the wires to the end posts with straining bolts and to the intermediates with staples. Do not drive the staples completely home, but allow the wires to run freely through them. The two lower wires (12 gauge/2.5mm) are set 40 and 55cm (16 and 22in) from the ground and the upper wires (14 gauge/2mm) 90cm, 1.2m and 1.5m (3, 4 and 5ft) (see p.57). Alternatively if both sides of each of the three sets of double wires can be taken down at the time of winter pruning, the task is made easier. To do this, fix a short length of chain to the ends of the wires and instead of wire staples on the intermediates, use small cup hooks. The chain is pulled tight around the end post so that the projecting end of the straining bolt is caught in one of the links. The wire itself is not attached to the bolt (see below).

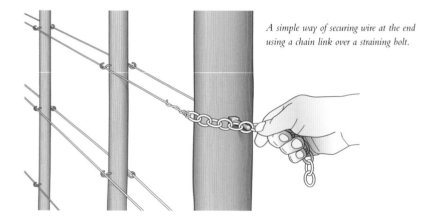

A simple way of securing wire at the end using a chain link over a straining bolt.

Spacing
Double Guyot

Space the vines 1.3–1.5m (4.5–5ft) apart in the row. Allow the wider spacing on good soils. The spacing between the rows is 1.5–1.8m (5–6ft). If a grass sward is preferred, then space the rows 1.8m (6ft) apart with a clean, grass-free band of 45cm (18in) wide maintained down the row.

Single Guyot
Space the vines 67–75cm (27–30in).

Planting

The young vines should each be planted to a stout cane 1.8m (6ft) in height. Plant to the same depth as the vine was in the nursery, being careful not to bury the graft union under the soil's surface where grafted plants are used. See 'Weed Control' (p.62) for details of planting through polythene.

Initial training

Normally, the vine will break into growth some time in May. In the first summer after planting allow the strongest shoot, usually the topmost, to grow and train it vertically up the cane, all others being cut out. In November, prune this shoot to a bud just below the bottom wire ensuring that at least two more good buds remain beneath it. In the second summer train in three shoots, tied vertically, pinching back any others to one leaf.

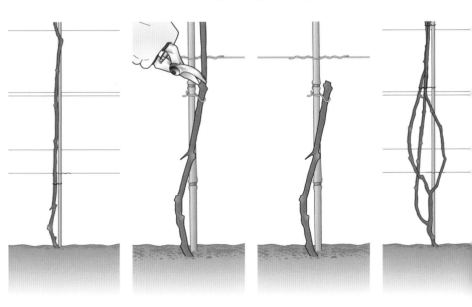

Left to right: a young vine at the end of the first summer, grown as a single stem; the young vine in autumn is cut to a bud just below the bottom wire; the young vine, with pruning completed; and the same vine in the autumn of the second year, when three shoots have been allowed to grow.

At the end of the second summer after leaf fall, start training to the Double Guyot system. One shoot is tied down to the left and one to the right on to the bottom wires. The third shoot is cut to three buds and this will provide the three replacement shoots for the next year (see photos p.60). The arms should not overlap with their neighbours but should be cut to a bud just short of the arms on the neighbouring vine.

A weak vine less than 60cm (24in) high at the end of the first year should be cut down to three buds and grown once again as a single-stemmed plant in the next growing season and then pruned as above.

Pruning the established vine

Refer to photographs on pages 60 and 61, also diagram opposite.

November. Prune as soon as possible after leaf fall, but no later than the end of January. However, early winter pruning stimulates earlier bud burst (by about seven days) than vines pruned in spring. Possibly where vines are planted in areas prone to spring frosts, there could be an advantage in pruning in March or April in the hope that they may escape frost damage, but it must be noted that late-pruned vines can bleed badly from the pruning cuts. Nevertheless, provided the vines are healthy, the bleeding does not appear to weaken the plants and the flow stops once the plant has developed some leaves. Note, there are, or should be, three strong replacement shoots trained up the cane or post. If not, utilize three strong laterals, nearest to the centre. The two arms that bore fruit last summer are cut back to the replacement shoots. One shoot is then tied down to the left and another to the right arching over the lower wires and the third is cut to three or four buds. These buds will provide the replacement shoots for next year.

January. Winter pruning completed. Replacement shoots arched over and tied to the lower wires and immature wood cut off, leave 60–75cm (24–30in) each side. Cut the third shoot to three or four buds.

July, August, September. The fruit-carrying laterals are trained through the double wires. Prune them to two or three leaves above the top wires as necessary, using secateurs or shears (see photo p.61). Remove any sub-laterals. The replacement shoots

Diagram of the pruning of a mature vine

Winter pruning

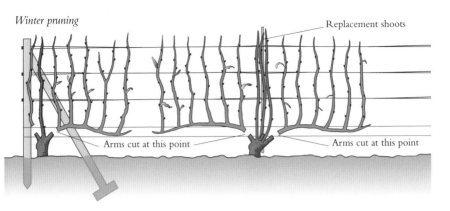

Replacement shoots

Arms cut at this point

Arms cut at this point

Winter: after pruning

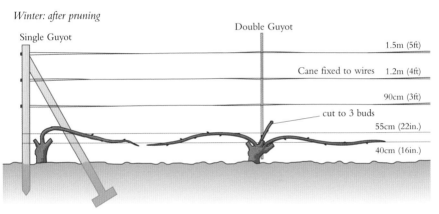

Single Guyot

Double Guyot

1.5m (5ft)

Cane fixed to wires   1.2m (4ft)

90cm (3ft)

cut to 3 buds

55cm (22in.)

40cm (16in.)

Summer (July, August, September)

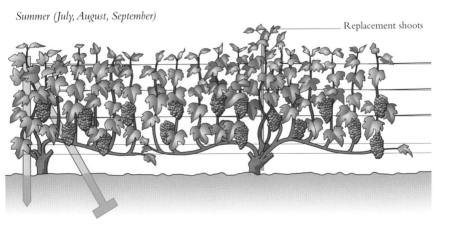

Replacement shoots

Top left: an established vine trained on the Double Guyot system before pruning.

Top right: remove arms back to young replacement shoots.

Centre left: after the arms are removed, a number of replacement shoots remain to be dealt with.

Bottom left: Three replacement shoots are left. One is arched over to the left and one to the right. The third is cut back to three buds, each of which will produce a replacement shoot next summer.

Bottom right: pruning is complete, all that remains to be done is to tie the arms to the lowest wires.

Top left: Double Guyot outside in spring.

Top right: Summer pruning a Double Guyot by cutting back the lateral shoots to two or three leaves above the topmost wire.

(three for a Double Guyot, two for a Single Guyot) are trained up the cane or post and stopped at about 1.5m (5ft). Sub-laterals are pinched to one leaf, and as a counsel of perfection, any blossom on the replacement shoots should be removed, though this is not strictly necessary if the growth is strong. Remove any suckers at the base and any surplus shoots growing from the main trunk below the two arms.

Starting in early September, gradually remove a small proportion of the lower leaves, just sufficient to expose the bunches to more sunlight and improve air circulation. Be careful not to expose the bunches too suddenly to the sun as this may lead to sun scorch. This reduction of foliage will help in the ripening of the berries and the control of botrytis.

Cropping

The vine should begin cropping in the third or fourth year depending upon the vigour of the plant. Allow only two or three bunches in the first year of fruiting, four or five in the next year and full cropping thereafter, provided that growth is healthy and vigorous enough to sustain the crop.

Feeding and watering

In February, apply Growmore fertilizer or equivalent at 66g/m² (2oz/sq yd) plus sulphate of potash at 15g/m² (½oz/sq yd) along each side of the row to a width of 30cm (12in) on either side. An organic alternative is seaweed meal applied at about 99g/m² (3oz/sq yd). On light soils in the spring, mulch the vines with a dressing of mushroom compost, bark, well-rotted manure or garden compost to a depth of about 5cm (2in). Vines are prone to magnesium deficiency, which is sometimes shown by yellowing of the leaves in summer (not to be confused with chlorosis due to alkaline conditions). If magnesium deficiency symptoms occur, apply a foliar spray of 220g magnesium sulphate (coarse Epsom salts) in 11 litres of water (½lb/3 gallons) plus a few drops of mild liquid detergent and repeat the spray 14 days later. Thereafter apply the magnesium sulphate as a top dressing in the late spring at 66g/m² (2oz/sq yd) (see also p.52).

Grapes grown for dessert will benefit from liquid feeds high in potassium in the growing season. Apply the fertilizer once a fortnight from the time the berries start to form until they begin to ripen.

Well-established vines are usually able to cope with dry weather except on very shallow soils; nevertheless, it is advisable to water them in drought conditions.

Ripening

The time of harvest will depend on the season of ripening of the particular cultivar (see p.64–71) and the weather. The grapes are ready when the sugar content within the berry is at its highest. On a professional basis one can use a refractometer to assess this, though most amateurs will probably have to rely purely upon the look and taste of the fruit. Cut off the whole bunch using secateurs. A handle is not necessary (see p.25) unless the grapes are wanted for dessert or exhibition. See also *Vines under Glass*, p.24–25.

Weed control

It is important to control weeds in the first year of planting when the vine is trying to become established, as well as in subsequent years when the plant is bearing a crop. Weeds compete for water and nutrients and are liable to create humid conditions around the bunches, making fruits more prone to grey mould. As the methods of weed control in vines are limited to hoeing or very careful use of the contact-action weedkiller mixture paraquat with diquat, it is essential to kill weeds regularly at intervals throughout the spring and summer months. This is necessary to prevent quick-developing annual weeds maturing and setting seeds, and to ensure that no perennial weeds have time to become established.

Paraquat with diquat will kill any green tissue, green bark, green buds or leaves on contact. It has a burning, scorching action and should be directed on to the weeds, but it must be kept clear of vines. It will kill annual weeds and check perennials, but the latter will usually survive to resume growth. Glyphosate is a non-selective herbicide and will kill any green plant material that it comes into contact with. Paint-on-gels can be used with care as a spot treatment to eliminate persistent perennial weeds.

Under no circumstances attempt to use the growth-regulator (hormone) type of weedkillers near vines as accidental contact of spray or spray drift can cause severe distortion.

Black polythene laid over the ground as a weed suppressant and as a mulch to conserve moisture is widely used by commercial growers and is worth adopting by the amateur. One can either use solid 500 gauge (125mm) , 1m (3ft) wide polythene, perforated or woven plastic. The latter two allow the rain to percolate down to the roots, but not the weeds to grow through. Lay the polythene along the row, cut slits or burn holes at each planting station and plant the vine through the aperture. The ground should be thoroughly prepared and in good condition as for normal planting and be well watered, not dry. Bury the sides of the polythene vertically about 15cm (6in) deep to prevent it blowing away. It is, of course, also possible to use plastic as a weed suppressant and mulch around vines already planted, but here two 50cm (20in) wide sheets are necessary, one on either side of the plants allowing about a 10cm (4in) overlap along the middle.

Cultivars for growing outside

Black grapes, other than a few hybrid cultivars, do not ripen sufficiently well when grown outside to make a wine of high quality, except perhaps in very hot summers. With one or two exceptions, they are best therefore grown on a wall or fence or given some protection for at least part of the growing season. For this reason in most commercial vineyards in Britain, white grapes predominate. In general, white cultivars that ripen in early to mid-October in the south are preferred as the early ones are liable to be attacked by wasps. Further north, early cultivars should be chosen rather than late ones, which might not ripen.

Cultivars noted as hybrids are not pure *Vitis vinifera* grapes but have one or more other *Vitis* species in their make-up. While they do not perhaps make as good a wine, they have great value in that most are highly resistant to powdery and downy mildew, thus reducing the need for spraying and probably eliminating it altogether.

Beside grapes for wine, there are a few outdoor cultivars also suitable for dessert in a good summer and these are listed below.

Nevertheless, as might be expected, grapes will acquire a sweeter flavour if grown against a warm wall. Cultivars normally recommended for under glass are unsuitable.

The season of ripening given below refers to the south of England.

WHITE

Bacchus (mid season) Similar to 'Müller Thurgau', but more reliable in cropping. Ripens its wood well. Gives a good quality wine.

White grapes need less heat to ripen than black grapes, therefore they can be grown in cooler climates.

Bianca (mid season) Hybrid with high resistance to powdery and downy mildew. Large bunches and berries of yellow grapes. Good cropper. Ripens mid-October.

Chardonnay (late season) A first-quality white wine and champagne grape that does best on light chalky loams. It is not suited to heavy clay. Yield moderate.

Chasselas (late season) There are a number of clones, e.g. 'Chasselas d'Or', 'Chasselas 1921', 'Chasselas Rose Royale' and 'Royal Muscadine'. All yield good berries excellent for dessert and wine. This cultivar is too late in the open and is better on walls. Late October.

Below: 'Chasselas Rose Royale'

Himrod Seedless (mid-season) Hybrid. Good for dessert, but needs a warm situation and is best on a wall. Berries are small and yellow. Flavour sweet, good. Large, loose bunches.

Huxelrebe (early mid-season) Vigorous. Heavy cropper and needs thinning on a young plant. Susceptible to wasps and botrytis. Produces a high-quality, muscat-flavoured wine.

Madeleine Angevine (early season) A vigorous, heavy cropper but prone to mildew. Relatively hardy and suitable for colder areas. Makes a white wine but quality only fair. Late September to early October.

Müller Thurgau (mid-season) Large bunches. Has a delicate flavour and gives an excellent wine but needs good weather during pollination, which means it can be an inconsistent cropper. Mid-October.

New York Muscat (mid- to late season) Hybrid. A tawny coloured grape of excellent muscat flavour in a hot summer. Must be grown on a wall. Large berries. Mid- to end October. Dessert.

Above:
'Muller Thurgau'

Orion (mid-season) Hybrid. Large, loose bunches with large berries. Very good cropper and highly resistant to powdery and downy mildew. Ripens mid-October.

Ortega (early season) Average cropper with medium-sized berries and bunches. An early ripener, beginning of September. Dessert and wine.

Perle de Czaba (early season) An excellent, muscat-flavoured berry. Best on a wall or under glass. Weak to moderate vigour. Light cropper. Late September. Dessert.

Perlette (early to mid-season) Small, yellow-green seedless berries of good flavour. A moderate cropper. Vigorous vine, needs a warm wall, ripens late September. Dessert and wine.

Phoenix (mid-season) Hybrid. Large, loose bunches and berries. Delicate muscat flavour. Highly resistant to powdery and downy mildew. Crops well and ripens mid-October. Dessert and wine.

Précoce de Malingre (early season) A large-berried sweetwater grape of good flavour. Moderate vigour. Light cropper. Dessert and wine. Late September.

Reichensteiner (early mid-season) Moderately vigorous. Crops well. Grapes high in sugar, but makes a neutral wine and best blended. Some resistance to botrytis.

Schönburger (late season) Moderately vigorous. Makes a good-quality wine but requires a warm, sheltered site to crop well.

Seyval (Seyve Villard 5.276) (mid-season) A hybrid with some resistance to powdery and downy mildew. A heavy cropper and easy to grow. Blends well with **'Müller Thurgau'**. Recommended. Mid- to end October.

Siegerrebe (very early season) A golden brown berry of good flavour with a trace of muscat and high sugar content. Prone to wasps. Dessert and wine. Medium vigour. Late August to early September.

'Siegerrebe', a very early grape, suitable for dessert and wine.

BLACK

Brant (late season) A hybrid of Canadian origin often grown for its autumn colour, bears heavy crops of small, sweet black grapes liked by children. Some resistance to mildew. Vigorous, useful wall cover. Mid-October.

'Cascade', a vigorous black grape that ripens in early October.

Left: 'Seyval'.

Cascade (Seibel 13.053) (mid-season) Resistant to powdery mildew. Low in sugar and high in acid. A fair-quality wine. Small bunches of deep purple berries with red juice. Very vigorous, excellent as a wall cover plant. Early October.

Muscat Bleu (mid-seaseon) A muscat-flavoured berry of good size. Bunches medium. Excellent flavour but seeds are large. Best on a wall, but will ripen out in the open during a warm summer. Attractive autumn colour. Mid-October. Dessert.

Noir Hatif de Marseilles (early season) Small to medium-sized berries and bunches of slight muscat flavour. Good quality. Rather a weak grower and requires fertile soil. Needs protection or a warm wall. Late September. Dessert.

Pinot Noir (early season) Similar to Pinot Noir (later season) but does not require the same warm conditions, e.g. wall, glass or polythene. Ripens early October.

Pinot Noir (late season) A very high quality grape used for champagne. In most years in Britain it needs the warmth of a wall, glass or a polythene tunnel to ripen the berries fully. The wood of the vine itself matures well and flower initiation is good, even in a cool summer.

Queen of Esther (early to mid-season) Hybrid. Good cropper, high resistance to powdery and downy mildew. Large loose bunches and berries. Excellent flavour. Ripens mid-September. Good autumn colour. Dessert and wine.

Regent (mid-season) Hybrid with high resistance to powdery and downy mildew. Good cropper. Medium to large bunches and berries. Good sweet flavour. Attractive autumn colour. Ripens early October. Dessert and wine.

Roem van Boskoop (Gloire de Boskoop or Boskoop Glory) (late season). A Dutch sweetwater black grape of good dessert quality but requires a warm wall, glass or polythene to ripen properly.

Rondo (mid-season) Hybrid. Good resistance to powdery and downy mildew. Heavy cropper. Medium to large berries and loose bunches. Sweet. Ripens early October. Dessert and wine.

'Tereshkova'.

Tereshkova (early season) A hybrid with medium-sized berries of purple-red with delicate muscat flavour. Attractive autumn colour. Late September. Dessert.

Triomphe d'Alsace (early) Very vigorous, crops heavily. A hybrid that is resistant to mildew. Makes an intensely coloured red wine of fair quality. Suitable outside and against a wall.

AMERICAN VINES

These are North American hybrids with the vine species *Vitis labrusca* in the parentage. There are a number of cultivars such as **'Suffolk Red'**, **'Glenora'** (blue/black), **'Venus'** (blue/bronze), **'Reliance'** (pink/red), **'Concord'** (black) and **'Fragola'** (pink). They have small or large bunches of good-sized berries, having a thick, slippery skin and a taste faintly reminiscent of strawberries, but can be an aqcuired taste. The grapes are sometimes described as having a foxy flavour, which is imparted into wine made from them. They are best grown against a wall and have a good resistance to mildew. Grapes ripen in late October.

PROPAGATION, PESTS AND DISEASES

The propagation of vines is quite simple as they have a tremendous propensity for rooting. The use of single buds enables the gardener to produce a large number of plants from a given length of one-year-old ripened growth.

Propagation takes place in late winter to early spring using winter prunings that have been inserted in a sheltered spot in open ground for the winter. For the single bud method, it is essential that a heated propagator is available. Make the top cut just above a bud, with the bottom cut about 5cm (2in) below so that there is a piece of longer stem at the base, which allows for the cuttings to be firmly inserted into small pots, with the bud just flush with the surface of any general purpose compost. When placed on bottom heat, rooting will soon take place so that within a month or two they can be potted on, ultimately finishing in a 15cm (6in) pot in the first season.

A cold greenhouse or garden frame can also be used but rooting will take longer when using the following method: short lengths of two buds of one-year-old growth inserted around the edge of 15cm (6in) pots filled with a gritty potting compost. Rooting will take several months but the cuttings should be ready to pot up by late summer. This will certainly apply to similar cuttings that have been prepared with three buds and rooted in the open ground. If the soil is naturally sandy and/or very well drained, all that needs to be done is to cultivate the ground to a good depth and insert the cuttings to two buds firming the soil afterwards. It will be necessary to water in dry conditions.

'Seyval' grown as a standard. Note the supporting stake. This hybrid shows some resistance to powdery and downy mildew.

Green softwood cuttings can also be made in July and August using shoots of the current year's growth. Cut lengths of stem about 10cm (4in) long just below a bud, to include either the growing tip or at least two buds. Insert the cutting into coarse sand or perlite using individual small pots and place in a propagating case. First move into a 9cm (3½in) pot when sufficiently rooted and subsequently into a 13cm (5in) pot. John Innes no. 2 potting compost will be ideal for these initial pottings. In the first year, the young vines, although hardy, will make better growth in a cool greenhouse.

Vines can also be grafted. Some commercial UK growers of wine grapes graft their vines on to rootstocks resistant to *Phylloxera vastatrix* (see p.78).

PESTS

Grapes grown out of doors are generally much less liable to be attacked by pests troublesome in greenhouses, such as mealybugs, scale insects, red spider mite and whitefly. However, an outdoor vine growing in a warm, sheltered position, such as against a south-facing wall, may also have these problems. When applying pesticides there is a danger that the bloom on glasshouse grapes will be marred if sprays are applied after the fruits have begun to swell. As far as possible, pests should be controlled before this stage is reached. Very few pesticides are available today for use on grape vines. Biological controls will deal with some pests, but it may sometimes be necessary to use organic pesticides to reduce infestations.

Some of the pests listed below have a wide range of host plants. If the vine is being grown in a greenhouse with other plants, these should also be checked for the presence of pests and treated accordingly.

Birds

Birds are the most damaging pests of outdoor grapes and they may eat the entire crop unless the vine is covered with birdproof netting. This will need to be in place from the beginning of September to the end of October depending on the cultivars grown. Netting with a mesh of 2–2.5cm (¾–1in) must be used. If the vine is too large to be netted completely,

Vines covered in netting to protect the crop from birds.

74

individual bunches can be enclosed in bags made from muslin or nylon tights (see p.79).

Glasshouse red spider mite

Red spider mites are tiny, eight-legged, sap-feeding animals that occur in large numbers on the underside of the leaves. They overwinter as adult mites and they begin feeding on the new foliage during April or May.

The first sign of an attack is a fine yellowish-green speckling that can be seen on the upper leaf surface; examination of the underside of the leaf with a hand lens should reveal the mites. As the infestation increases, leaves begin to dry up and drop off, and the vine may become covered in a fine silken webbing, which is produced by the mites. Despite their common name, the mites are not really red, and they range in colour from nearly black to yellowish-green or orange.

Early treatment is required to prevent a damaging infestation from developing. As soon as mite damage is seen, introduce a predatory mite called *Phytoseiulus persimilis*, although this cannot be used if insecticides are being used to control other pests. Supplies can be obtained from mail order suppliers of biological controls or ordered through some garden centres. The predator works best between late March and early October. If spider mite control is needed at other times, spray with vegetable oils or rotenone/derris.

Glasshouse whitefly

Adult whiteflies are tiny, moth-like insects with white wings, and they live on the underside of the leaves. The nymphal stages are flat, whitish-green scales, and both the adults and nymphs feed by sucking sap. Like mealybugs, whitefly can create problems with honeydew and sooty mould. The eggs and nymphal stages are relatively immune to insecticides, so frequent sprays are necessary to kill the adults as they develop. Sprays containing vegetable oils or pyrethrum can reduce infestations, provided resistance has not occurred. As an alternative to insecticides, a parasitic wasp called *Encarsia formosa* can be introduced during the summer. This is available from the same sources as *Phytoseiulus* (see above).

Mealybug

This is a common pest of glasshouse grapes. Mealybugs are soft-bodied, pinkish-grey insects up to 3mm (⅛in) long, and they suck sap from the leaves and stems. They tend to cluster in the leaf axils and they secrete white woolly fibres around themselves. They excrete a sugary substance called honeydew, which makes the vine sticky and encourages the growth of a black sooty mould on the surface of the foliage and fruits.

A ladybird beetle, *Cryptolaemus montrouzieri*, which preys on mealybugs and their eggs, can be obtained from mail order suppliers of biological controls or ordered through some garden centres. Mealybugs are difficult to control, especially once they get among the developing fruits. The ladybird predator needs high temperatures and so it is unlikely to give good control early in the season. Spraying with vegetable oils may help to reduce infestations.

Scale insects

Scale insects feed by sucking sap from the vine and the type most commonly found is called brown scale. The mature insects are covered by hard, shiny brown, convex shells about 6mm (¼in) long, and they are attached to vine stems. Apart from the recently hatched nymphs, scale insects do not move once they have found a suitable feeding place.

A less common but more spectacular type of scale insect is the woolly vine scale. The mature females of this scale have dark brown, wrinkled shells about 4mm (³⁄₁₆in) in diameter, which are perched on the edge of a mass of white waxy threads, which contain the eggs.

During the winter, the vine rods should be scraped to remove loose bark and as many scales as possible. Collect and burn the scrapings. The chemicals available to amateur gardeners are not good for scale control on vines, but spraying with vegetable oils in late June and mid-July, as the young nymphs begin to hatch, may reduce attacks.

Vine blister mite (*Erinose*)

This is a microscopic mite that can occur on both glasshouse and outdoor grapes. It overwinters inside the buds and starts feeding on the leaves in the spring. As a result of this feeding, the upper surfaces of the leaves develop raised patches and the undersides of these blisters are covered by fine white hairs. These hairs may be confused with mildew disease, but the blistering effect is diagnostic for the mites. The symptoms first appear in May or June, and as the summer progresses the hairs darken and become reddish-brown. The mites live among the hairs and feed on the leaf surface but, apart from some disfigurement to the leaves, they seem to cause no real harm. Infestations can usually be dealt with by picking off the affected leaves, but not if this would result in significant defoliation.

Vine phylloxera

This serious pest of vines is rare in this country, although outbreaks occasionally occur when infested vines are imported into Britain. It is an aphid-like insect with a complex life cycle, having different forms that cause galls on the leaves and roots. The leaf galls are the most obvious symptom. They are pinkish or yellowish-green, swollen, rounded structures, which form on the leaf surface. The galls are hollow and contain the pest. The root-infesting forms cause swellings on the roots and heavy infestations will kill the plant unless the vine has been grafted on to a phylloxera-resistant rootstock. Suspected outbreaks must be reported immediately to the local office of the Department for the Environment, Food and Rural Affairs (DEFRA) who will take the appropriate control measures.

Vine weevil

The adult weevils feed at night and eat irregular-shaped notches from the leaf margins of many plants, including grape vines. The beetles are mainly black with small brown patches on the wing cases. The slow moving beetles are about 9mm (⅜in) long and they have jointed antennae. Their larvae are fat, creamy white, legless grubs with brown heads. They live in the soil and feed on roots. Well-established vines are unlikely to suffer serious damage, but pot-grown or young vines may be killed by the grubs. If damage caused by the grubs or adults is seen, they can be controlled by drenching the soil in the summer with a pathogenic nematode, *Steinernema kraussei*. The nematode is available from mail order suppliers of biological controls or can be ordered through some garden centres.

Wasps

In some years wasps are very numerous and they will damage large numbers of grapes, especially of the early ripening cultivars. An effort should be made to find and destroy as many wasp nests as possible in the locality. It is easier to follow the wasps' flight paths as the sun is setting and likely places to look for nests are ditches, hedge bottoms, and under the eaves of buildings. Once located, the nests can be destroyed by placing a wasp powder such as bendiocarb in the entrance hole. Individual bunches of grapes can be protected from wasps by enclosing them in bags made from muslin or old nylon tights. Wasps can also be kept out of greenhouses by screening the ventilators with nylon netting.

The fruit can be protected from wasps and birds by an old nylon stocking.

DISEASES

Downy mildew

Downy mildew occurs only occasionally in this country, mainly on outdoor grapes. It shows as light green blotches on the upper surface of the leaves and as a downy mildew on the lower surface of these patches. Affected areas dry up and become brittle causing the leaves to curl and fall. Diseased berries shrivel and become brown and leathery. The tips of the shoots may also be attacked. Remove and destroy all diseased tendrils and leaves to remove the overwintering stage of the fungus, although some spores also overwinter in the bud scales and on shoots. There are no fungicides available to amateur gardeners for control of this disease.

Berries badly affected by grey mould.

Grey mould

Grey mould (*Botrytis*) is the most troublesome disease that occurs on outdoor vines, but it can also be a nuisance on indoor grapes in poorly ventilated greehouses where humidity is high.

Affected grapes rot and become covered with a dense brownish-grey furry mass of fungal growth. The fungus either attacks the berries through wounds or it may invade the floral parts so that some of the fruits are already diseased as they develop. By whichever of these methods infection occurs, once the disease is established it can spread rapidly by contact and also by means of air-borne spores, which are produced in vast numbers. In wet weather and in very humid greenhouses crop loss can be considerable. Under glass the trouble can be prevented to a certain extent by adequate ventilation to reduce the humidity, and by the prompt removal of unhealthy berries and leaves.

It is difficult to control the disease on outdoor grapes so in wet seasons try to improve the aeration around the bunches by thinning, and by the judicious removal of some of the shoots or leaves. There are no fungicides available to amateur gardeners for control of this disease. For those growing grapes on a commercial scale, fungicides are available to control this disease, and advice should be sought on this subject.

Honey fungus

Both indoor and outdoor vines are very susceptible to infection by this soil-borne fungus, which can kill affected plants very rapidly. White fan-shaped growths of fungus develop beneath the bark of the roots and main stems at and just above ground level. Brownish-black root-like structures known as rhizomorphs may be found growing on diseased roots; these grow out through the soil and spread the disease. Control of the fungus is difficult and it is essential, therefore, to trace the source of infection so that all woody debris can be dug up and burnt together with dead and dying vines and as many roots as possible. (For more detailed advice, consult the Advisory Service at RHS Garden Wisley).

Powdery mildew

Powdery mildew is most troublesome on indoor vines , particularly in a cold greenhouse especially if the soil is dry and the atmosphere is humid or stagnant. Powdery mildew is also common on outdoor grapes, particularly on those growing in very dry positions such as against walls. A white powdery coating of fungus spores develops only sparsely and the most obvious symptom is a grey or purplish discoloration of the diseased areas. The disease can also attack the flowers and fruits causing them to drop, or at a later stage the grapes may become hard, distorted and split, and are then often affected by secondary fungi, such as grey mould (see p.80), which cause extensive rotting.

Powdery mildew can be prevented to a certain extent by mulching and watering to stop the soil from drying out. As soon as the disease appears apply sulphur dust. If the disease has been troublesome in previous years the first application should be given 10 to 14 days before mildew is expected. Four applications during the season may be needed to keep the disease in check. In a cold greenhouse in dull weather, it may be necessary to provide some heat temporarily to avoid too humid an atmosphere and careful but sufficient ventilation should be given to get good air circulation. Over-crowding of the shoots and leaves must be avoided to prevent stagnant air conditions.

DISORDERS

'Bleeding' of the wounds

Often as a result of late winter pruning, the sap flows freely from the pruning cuts, especially if the wounds are large. The sight of such 'bleeding' can be disconcerting to the inexperienced grower. There is no need for concern, however, as provided the vine is healthy and well established, the sap loss does not appear to harm the plant and it will eventually stop in the spring when the buds break into growth. There have been many remedies advocated in the past to staunch the flow, such as cauterising the wound, applying sealing wax, carpenter's knotting etc., but these rarely work once the 'bleeding' has started. The real answer is not to prune late, but to complete the job by the end of December, and, preferably, immediately after leaf fall.

Bleeding from a pruned shoot due to late winter pruning.

Exudations

A common phenomenon in the spring is the presence of transparent globules resembling eggs, on the lower leaf surfaces and petioles. These small round greenish or colourless droplets are not the eggs of any pest but are due to a natural exudation from the plant. This type of exudation commonly occurs on young growth and indicates that the root action is very vigorous and the plant is in good health. The symptoms are inclined to be more obvious, however, on plants growing under glass where the atmosphere is humid.

Fruit splitting

This trouble most commonly occurs as a result of an attack by powdery mildew (see p.81). However it is occasionally due to irregular growth such as may occur when heavy rain follows a period of drought. Affected grapes usually rot as a result of subsequent grey mould infection. There is very little that can be done to prevent this trouble apart from mulching to conserve moisture and by watering in dry periods before the soil dries out completely.

Magnesium deficiency

Vines are soon affected by a soil deficiency of magnesium, symptoms being produced in the leaves. These usually show as a yellowish-orange discoloration between the veins, but in some cultivars the leaves may have purplish blotches. Later the affected areas turn brown; these symptoms should not be confused with sun scorch (p.84). The deficiency can be corrected by spraying the foliage at the first signs of trouble, with 220g (½lb) of magnesium sulphate in 9.5 litres (2½ gal) of water plus a spreader, such as soft soap or a proprietary product. Apply two or three times at fortnightly intervals. Soil applications act more slowly, but are useful as a long-term cure. For outdoor grapes apply the magnesium sulphate in subsequent years as a soil dressing in late spring at $66g/m^2$ (2oz per sq yd).

Oedema

Oedema is a physiological disorder that is more likely to occur on vines grown under glass. Fairly late in the season, small wart-like outgrowths appear on the stalks when the fruits are developing and sometimes on the grapes themselves and even on the lower leaf surfaces. These outgrowths may break open and then have a blister-like or white powdery appearance, or they may become rusty coloured and appear as brown scaly patches. The trouble is caused when the roots of an affected plant take up more water than the leaves can transpire and this may be due to extremely moist conditions either in the soil or in the atmosphere or both. Once oedema has occurred, do not remove the affected parts as this will only make the trouble worse. No special treatment can be given and the remedy is to maintain drier conditions both in the air and soil; with correct cultural treatment the affected plant should recover in due course.

Scald and scorch

Scald, which shows as discoloured sunken patches on the berries, and scorching of the foliage, which results in large pale brown patches, is due to hot sun striking through glass on to moist leaf tissues. These troubles usually occur where the ventilation has been poor, but light shading may also help to prevent them. Remove affected berries and leaves.

Shanking

Shanking first shows at the early ripening stage when odd berries or small groups of berries do not colour and develop naturally. It starts as a dark spot along the stalk of the grape, which it finally girdles so that it shrivels and the grape stops developing and fails to ripen. The grapes are watery; black grapes turn red and white grapes remain translucent. The fruit, if tasted, will be found sour and unpalatable.

This disorder is usually a result of one or more unsuitable cultural conditions, or the penetration of stagnant soil by the roots, or over-cropping of the vine, which puts an undue strain upon the root system. The remedy, therefore, is to study the soil conditions to see whether the roots could have been affected by drought or waterlogging. Mulching or resoiling may be necessary if there is any sign of such damage. At the same time reduce the crop for a year or two until the vine has regained its vigour. When shanking occurs fairly early in the season it is sometimes possible to save the rest of the crop by cutting out the withered berries and spraying the foliage with a foliar feed, providing the vine is not being overcropped by taking too many bunches.

Spray damage

Damage by sprays, usually due to the misuse of weedkillers (such as 2, 4-D; mecopropp), occurs on both indoor and outdoor vines. Affected leaves become narrow and fan-shaped, are frequently cupped and the shoots twist spirally. Affected plants usually grow out of the symptoms in due course, but it is better to avoid such damage by the more careful use of hormone weedkillers. Keep special equipment for their use, do not spray on a windy day nor use within 360m (400 yards) of outdoor grapes. Close all ventilators in greenhouses if any spraying is to be carried out in the vicinity. Do not leave weedkillers in a greenhouse as vapours from them may affect plants on a hot day.

FURTHER INFORMATION

SOCIETIES SPECIALISING IN GRAPES AND OTHER FRUIT

California Rare Fruit Growers, The Fullerton Arboretum, California State University, Fullerton, California 92634 USA
National Boomgarden Stichting (NBS), Informatiecentrum en Secretariat,Kersendaal 1, 3724 Vliermaal, Belgium
Tel 012 39 11 88 Email: nbs@wanadoo.be
RHS Fruit Group, RHS Garden Wisley, Woking, Surrey GU23 6QB
Tel: (01483) 224234
United Kingdom Vineyards Association, Church Road, Bruisyard, Saxmundham, Suffolk IP17 2EF
Tel: (01728) 638080

UK GARDENS TO VISIT

Brogdale, Brogdale Road, Faversham, Kent ME13 8XZ
Tel: (01795)531710
Hampton Court Palace, Surrey KT8 9AU
Tel: (020) 8781 9500
Groombridge Place, Groombridge, Royal Tunbridge Wells, Kent TN3 9QG Tel: (01892) 863999
RHS Garden Wisley, Woking Surrey GU23 6QB
Tel: (01483) 224234
West Dean gardens, Chichester, PO18 0QZ
Tel:(01243) 811342

'Seyval' grows well outside in the UK, giving a heavy crop from mid- to late October.

UK NURSERIES THAT SPECIALISE IN VINES

Deacon's Nursery, Moor View, Godshill, Isle of Wight P038 3HW. Tel:(01983) 840750 Email: deacons.nursery @btopenworld.com Website:**www.deaconsnurseryfruits.co.uk**

Highfield Nurseries, Whitminster, Gloucester GL2 7PL Tel: (01452) 740266/741309

Frank P. Matthew Ltd. Berrington Court, Tenbury Wells, Worcestershire WR15 8TH Tel (01584) 810214. Wholesale suppliers to independent garden centres only.

Reads Nursery, Hales Hall, Loddon, Norfolk NR14 6QW Tel: (01508) 548395 Email: plants@readsnursery.co.uk Website:**www.readsnursery.co.uk**

Scotts Nurseries (Merriott) Limited, Merriott, Somerset TA16 5PL Tel:(01460) 72306

Clive Simms, 'Woodhurst', Essendine, Stamford, Lincolnshire PE9 4LQ Tel:(01780) 755615 Email: clive_simms@lineone.net Website:**www.clivesimms.com**

Sunnybank Vine Nurseries, The Old Trout Inn, Dulas, Herefordshire HR2 0HL Tel: (01981) 240256 Email: vinenursery@hotmail.com Website:**www.vinenursery.netfirms.com**

Vigo Ltd, Dunkeswell, Honiton, Devon, EX14 4LF Tel: (01404) 890262, Email: sales@vigoltd.com Minimum order 25 vines

The Vine House, 3 Elm Street, Skelmanthorpe, Huddersfield, West Yorkshire HD8 9BH Tel: (01484) 865964 Email:sales@thevinehouse.co.uk Website:**www.thevinehouse.co.uk**

Vines and Olives in Pots, Jane Phillips, Fosseway, Hinton St. George,Somerset TA17 8SE Tel: (01460) 72420

US NURSERIES THAT SPECIALISE IN VINES.

Miller Nurseries, 5060 West Lake Road, Canandaigua, NY 14424-8904 Tel:1-800-836-9630

St. Francois Vineyards, 1669 Pine Ridge Trail, Park Hills, MO 63601 Tel/Fax:(573) 431-4294

Double A Vineyards, 10277 Christy Road, Fredonia, NY 14063 Tel: (716) 672-8493

BOOKS

Growing Grapes in Britain - a handbook for winemakers, Gillian Pearkes (Amateur Winemaker, 1969)

Growing Vines to Make Wine, Nick Poulter (Nexus Special Interests, 1998)

Growing Wine Grapes, J.R. McGrew, A. Hunt, T. Zabadal (G.W. Kent, 1994)

Successful Grape Growing for Eating and Winemaking, S. Read (Groundnut Publishing, 2001)

The Cultivated Fruits of Britain, F.A.Roach (Blackwell, 1985)

Vinegrowing in Britain, Gillian Pearkes (J.M. Dent & Sons, 1982)

Vines and Vine Culture, A.E. Barron (London Journals of Horticulture,1912)

Vines and Wines in a Small Garden - from planting to bottling, James Page-Roberts (The Herbert Press, 1995)

WEBSITES

www.agardenstore.com/grape.html
Basic advice for beginners
www.englishwineproducers.com
all the latest news and information about English wines, vineyards and producers
www.grapeseek.com
a search engine and directory that focuses on grapes, winemaking, and related topics
www.grapepictures.com
access to a range of photographs of grape varieties, vineyards and techniques
www.wineloverspage.com/wineguest/wgg.html
A large grape glossary

INDEX

Page numbers in **bold** refer to illustrations

'Alicante' 21, 40, **40**
American vines 71
arches 46
artificial heat 10, 12, 19
ash 13
associations, fruit specialists 86
atmosphere, greenhouses 19, 21, 24

'Bacchus' 64
bark removal 18, 77
'Bianca' 66
biological controls 74
    mealybugs 77
    red spider mites 76
    vine weevils 78–9
    whitefly 76
birds 74–6
'Black Hamburgh' 21, 24, 26, **27**, **31**, 37, 40
black polythene 63
bleeding 15, 17, 58, 82, **82**
blisters 78
bloom 21–2
books 89
'Boskoop Glory' 71
Botrytis (grey mould) 80–1, **80**
'Brant' 52, 69
brown scale 77
'Buckland Sweetwater' 26, 37, 40
bunches
    handles 24
    shaping 22–3
    thinning 21, 52

'Canon Hall Muscat' 40–1
'Cascade' **2**, 52, 69, **69**
chalk 53, 54

'Chardonnay' 66
'Chasselas' clones 52, 66
'Chasselas Rose Royale' **66**
clay 53, 54
climate 44
compost 13
'Concord' 71
construction, ground vineries 34–6
cordons 8, **50**
cropping, vineyard plants 61
Cryptolaemus montrouzieri 77
cultivars
    see also hybrids; *Vitis vinifera*
    ground vineries 37
    growing under glass 38–43
    outdoor growing 64–71
    pot-grown 26
'Curate's Vinery' 34
cuttings 72, 74

dessert grapes 10–12
    feeding 52, 62
    thinning 52
diseases 80–2
    Botrytis 80–1, **80**
    downy mildew 64, 80
    grey mould 80–1, **80**
    honey fungus 81
    powdery mildew 81–2
disorders 82–5
    bleeding 82, **82**
    exudations 83
    fruit splitting 24, 83
    magnesium deficiency 83
    oedema 84
    scald and scorch 84
    shanking 84–5
    spray damage 85
Double Guyot 55, 56
    see also Guyot system
downy mildew 64, 80
    see also mildew
drainage 13, 53
drought 51, 62

Encarsia formosa 76
Epsom salts 62
Erinose (vine blister mite)
    77–8
espaliers 8
exudations 83
eyes **9**

feeding
    see also fertilizers
    greenhouse vines 23
    outdoor vines 52
    pot-grown vines 29
    vineyard plants 62
fences 46–7
    vineyards 56–7, **56**
fertilizers
    see also feeding
    greenhouse vines 18–19
    soil preparation 13, 48, 54
flowering, greenhouse vines
    20–1
'Foster's Seedling' 21, 24, 26,
    37, 41
'Fragola' 71
'Frontignan' 41
'Frontignan' types 26
frost 53, 58
fruit
    splitting 24, 83
    thinning 21–3
fungicides 80, 81

gardens 86
glass
    cultivars 38-43
    greenhouses 10–25
    ground vinery 34–7
glasshouse red spider mites 76
glasshouse whitefly 76
glazing materials 34
'Glenora' 71
'Gloire de Boskoop' 71
glossary 8–9

glyphospate 63
'Golden Chasselas' 41
'Golden Queen' 41
graft union 54
grafted vines 54, 74
greenhouses 10–25
    diseases 80–2
    heat 10, 12, 19
    pests 18, -76–9
    planting 14–15, **14, 15**
    shading 20
    suitable cultivars 38–43
    temperature 19, 24
    wires 14
grey mould (*Botrytis*) 80–1, **80**
'Gros Colmar' 41
ground vineries 34–7, **35, 36**
growth-regulator weedkillers
    63
Guyot system 55–61, **57, 59,
    60, 61**

handles, bunches 24
harvest time 12, 24, 62
'Himrod Seedless' 66
honey fungus 81
honeydew 76, 77
hormone weedkillers 63, 85
humidity
    greenhouses 19, 21, 24
    grey mould factor 80–1
'Huxelrebe' 66
hybrids
    see also cultivars; *Vitis
        vinifera*
    North American 71
    'Bianca' 66
    'Brant' 69
    'Himrod Seedless' 66
    'New York Muscat' 43, 52,
        67
    'Orion' 67
    'Phoenix' 67
    'Queen of Esther' 70, **70**
    'Regent' 70–1

'Rondo' 71
'Seyval' 7, **68**, **68**, **73**, **87**
'Tereshkova' 71, **71**
'Triomphe d'Alsace' 71

John Innes potting compost
no.3 13

keeping, grapes 25

'Lady Downe's Seedling' 25,41
'Lady Hutt' **39**, 42
laterals 8, **9**
    lowering **19**, 20
    pot-grown vines 29
    thinning 50
layering 31–3
leaf galls 78
leaves
    removal 61
    yellowing 62

'Madeleine Angevine' **49**, 66
'Madresfield Court' 42
magnesium deficiency 62, 83
manure 13, 18
mealybugs 18, 77
mildew 80, 81–2
    outdoor vines 51
    resistance 64
    wall-trained vines 46
'Mrs Pearson' **11**, 42
'Mrs Pince' 24, 42
mulching
    grafted vines 54
    indoor beds 18
    outdoor vines 48, 51
    powdery mildew prevention
        81
    vineyard plants 62
'Müller Thurgau' 45, 67, **67**
multiple cordons 8, **9**, **47**

Muscat grapes 21, 38, 40–3
'Muscat of Alexandria' 24, **25**,
    42, 43
'Muscat Bleu' 69
'Muscat Champion' 43
'Muscat Hamburgh' 43

nematodes 78–9
netting, bird protection 74, **75**,
    76
'New York Muscat' 43, 52, 67
'Noir Hatif de Marseilles' **65**,
    69
nurseries 88–9

oedema 84
'Orion' 67
'Ortega' 67
outdoor vines 44–63
    see also vineyards
    feeding 52
    planting 48
    pruning 48–51
    revival 8
    thinning 52
    training 46–7
outside planting, greenhouse
    vines 12, 14–15
overcropping 52, 84

paraquat with diquat 63
parasitic wasp 76
pergolas 46–7
'Perle de Czaba' 67
'Perlette' 67
pesticides 74
pests 74–9
    birds 74–6
    brown scale 77
    control 18
    mealybugs 77
    red spider mites 76
    scale insects 77

vine blister mites 77–8
vine louse (*Phylloxera*) 54
vine weevils 78–9
wall-trained vines 46
wasps 79
whitefly 76
woolly vine scale 77
'Phoenix' 67
Phylloxera vastatrix (vine
   louse) 54
Phytoseiulus persimilis 76
pinching back 9, 19–21, **19, 20**
   outdoor vines 50
   pot-grown vines 29–30
   sub-laterals 61
'Pinot Noir' 70, **70**
planting
   see also soil
   depth 54
   greenhouse vines 12, 14–15,
      14, 15
   ground vineries 37
   outdoor vines 48
   polythene sheets 63
   vineyards 54, 57
pollination, greenhouse vines
   21
polythene sheets 63
posts, Guyot system 55–6
pot-grown vines 10, 26–33
   cultivars 26, 38–43
   methods 28–33, **30–3**
   training 28–9, **28, 29**
powdery mildew 64, 81–2
   see also mildew
'Précoce de Malingre' 52, 67
preparation, soil 13, 48, 53–4
propagation 72–4
protection
   birds 74–6, **75**
   wasps **79**
   weather 44, 46
pruning
   established vines 17
   first years 16–17
   ground vineries 37

Guyot method 57–61, **57,
   59, 60, 61**
   outdoor vines 48–51, **51**
   planting time 15
   pot-grown vines 28–9
   replacement method 17,
      57–61, **57, 59, 60, 61**
   rod and spur system 9,
      16–17, **16**
   summer 16, 23–4, 49, 50,
      58, 61, **61**
   winter 17, 50–1, 58, **59**
pyrethrum 76

'Queen of Esther' 70, **70**

rain shadow 46
rainfall 44
'Red Flame Seedless' 43
red spider mites 18, 46, 76
'Regent' 70–1
'Reichensteiner' 68
'Reliance' 71
renovation, inside beds 18
replacement pruning 17
   see also Guyot system
repotting 28, 29
resistance, mildew 64
ripening
   dessert grapes 10
   greenhouse grapes 23–5
   growth 17, 26, 29
   pot-grown grapes 29
   vineyard grapes 61, 62
rod 9
rod and spur system 9, **9,**
   16–17, **16,** 29
'Roem van Boskoop' 71
'Rondo' 71
roots 53
   cuttings 72
   greenhouse vines 12
   pot-grown vines 31–2
rootstocks 54

sap, bleeding 15, 17, 58, 82, **82**
scald 84
scale insects 77
'Schönburger' 68
scorch 61, 84
'Seyval' **7**, 68, **68**, **73**, **87**
shading
   greenhouses 20
   ground vineries 37
   scorch prevention 84
shanking 84–5
shaping, bunches 22–3
shoots **28**
   pinching 19–21, **19**
   stopped **20**
   summer pruning 23–4
   surplus **19**
shoulders 22–3
'Siegerrebe' 52, 68, **68**
single bud method,
propagation 72
Single Guyot 55, 57
   see also Guyot system
site selection, vineyards 53
societies 86
softwood cuttings 74
soil
   conditions 84–5
   preparation 13, 48, 53–4
   renovation 18
   requirements 53
sooty mould 76, 77
spacing
   cordons 55
   Guyot system 55, 56–7
   rods 46
   vineyard plants 56–7
splitting 24, 83
spraying
   water 19, 31
   weedkillers 85
spring, greenhouse vines
   19–20
spurs 9, **51**
standards, pot-grown vines 26,
   28–9, **28**

Steinernema kraussei 78–9
stopping *see* pinching
sub-laterals 9, **20**, 61
suckers 61
'Suffolk Red' 71
sulphur dust 82
summer pruning
   greenhouse vines 16
   Guyot method 58, 61, **61**
   outdoor vines 49, 50
support
   see also wires
   outdoor vines 47
   vineyard plants 55–6
Sweetwater grapes 38, 40–3

temperature, greenhouses 19,
   24
tendrils 9, 20, 24
'Tereshkova' 71, **71**
thinning
   bunches, outdoor vines 52
   grapes 21–3, **22**, 52
   laterals 50
'Thompson Seedless' 43
top dressing 18, 29
training
   greenhouses vines 14
   ground vineries 37
   outdoor vines 46, 48–9
   pot-grown vines 28–9
   standards 28–9
   vineyards 55–61, **57**
'Triomphe d'Alsace' 71
tying in
   outdoor vines 50
   standards 29, **29**

vegetable oils, pest control 76,
   77
ventilation
   greenhouses 10, 19, 21
   grey mould prevention
      80–1

ground vineries 34, 36, 37
'Venus' 71
vine blister mite (*Erinose*)
77–8
vine louse (*Phylloxera vastatrix*)
54
vine weevils 78–9
vineyards 53–63
  cropping 61
  cultivars 64
  feeding 62
  planting 54, 57
  pruning 57–61, **57, 59, 60,
    61**
  site selection 53
  soils 53–4
  supports **56**
  training 55–61
  watering 62
  weed control 62–3
  wires 56, **56**
Vinous grapes 38, 40–1
Vitis vinifera
  see also cultivars; hybrids
  climatic needs 44
  origins 6
'Alicante' 21, 40, **40**
'Bacchus' 64
'Black Hamburgh' 21, 24,
  26, **27, 31**, 37, 40
'Boskoop Glory' 71
'Brant' 52
'Buckland Sweetwater' 26,
  37, 40
'Canon Hall Muscat' 40–1
'Cascade' **2**, 52, 69, **69**
'Chardonnay' 66
'Chasselas' clones 52, 66
'Chasselas Rose Royale' **66**
'Foster's Seedling' 21, 24,
  26, 37, 41
'Frontignan' 41
'Frontignan' types 26
'Gloire de Boskoop' 71
'Golden Chasselas' 41
'Golden Queen' 41

'Gros Colmar' 41
'Huxelrebe' 66
'Lady Downe's Seedling'
  25, 41
'Lady Hutt' **39**, 42
'Madeleine Angevine' **49**,
  66
'Madresfield Court' 42
'Mrs Pearson' **11**, 42
'Mrs Pince' 24, 42
'Müller Thurgau' **45**, 67, **67**
'Muscat of Alexandria' 24,
  25, 42, 43
'Muscat Bleu' 69
'Muscat Champion' 43
'Muscat Hamburgh' 43
'Noir Hatif de Marseilles'
  65, 69
'Ortega' 67
'Perle de Czaba' 67
'Perlette' 67
'Pinot Noir' 70, **70**
'Précoce de Malingre' 52,
  67
'Red Flame Seedless' 43
'Reichensteiner' 68
'Roem van Boskoop' 71
'Schönburger' 68
'Siegerrebe' 52, 68, **68**
'Thompson Seedless' 43

wall-trained vines 46–7
  see also outdoor vines
  cultivars 64
  pests 74
  planting 48
  watering 51
  wires 47
wasps 76, 79
watering
  greenhouse vines 18, 23
  outdoor vines 51
  powdery mildew prevention
    81
  vineyard plants 62

websites 89
weed control, vineyards 62–3
weedkillers 85
whitefly 76
wine grapes, thinning 52
winter maintenance, greenhouse
    vines 18–19
winter pruning
    greenhouse vines 17
    Guyot method 58, **59**
    outdoor vines 50–1
wires
    greenhouses 14
    ground vineries 37
    pergolas 47
    vineyards 56, 56
    wall-trained vines 47
wood ash 13
woolly vine scale 77

yellowing leaves 62

The Publisher would like to thank the following people for their kind permission to reproduce their photographs:

Harry Baker: p.2, 7, 25, 30 (left and right), 39, 45, 47 (top and bottom), 49, 50, 51, 60 (all), 61 (top and bottom), 65, 67, 68 (top and bottom), 69, 72, 75, 79, 82, 87
Nina Gregson: p.27, 34, 36
Nick Dunn/Frank P. Matthews Ltd.: p.70
Garden Picture Library: p.40 (Howard Rice), 70 (Tim Spence)
ABPL: p.66 (Anthony Blake), 81 (Georgia Glynn Smith)
Photos Horticultural: p.42
Ray Waite: p.11, 14, 15, 31 (left and right), 32 (top and bottom), 33

Jacket image: **Photos Horticultural**